DIFFICULT TO TREAT DEPRESSION

FIRST EDITION

Chris Aiken, MD

Editor in Chief
The Carlat Psychiatry Report

Director
Psych Partners

Adjunct Assistant Professor
Department of Psychiatry
New York University Langone
Wake Forest University

Published by Carlat Publishing, LLC
PO Box 626, Newburyport, MA 01950

Difficult to Treat Depression
FIRST EDITION

Published by Carlat Publishing, LLC
PO Box 626, Newburyport, MA 01950

Publisher and Editor-in-Chief: Daniel J. Carlat, MD
Deputy Editor: Talia Puzantian, PharmD, BCPP
Senior Editor: Ilana Fogelson

This CME/CE activity is intended for psychiatrists, psychiatric nurses, psychologists, and other health care professionals with an interest in mental health. The Carlat CME Institute is accredited by the Accreditation Council for Continuing Medical Education to provide continuing medical education for physicians. Carlat CME Institute maintains responsibility for this program and its content. Carlat CME Institute designates this enduring material educational activity for a maximum of twelve [12] AMA PRA Category 1 Credits™. Physicians or psychologists should claim credit commensurate only with the extent of their participation in the activity. The American Board of Psychiatry and Neurology has reviewed the Difficult to Treat Depression and has approved this program as part of a comprehensive Self-Assessment and CME Program, which is mandated by ABMS as a necessary component of maintenance of certification. CME tests must be taken online at www.thecarlatreport.com/cme (for ABPN SA course subscribers).

To order, visit www.thecarlatreport.com
or call (866) 348-9279

Print: 979-8-9998818-1-6
eBook: 979-8-9893264-6-4

PRINTED IN THE UNITED STATES OF AMERICA

To Kellie Newsome

Special thanks to Daniel Carlat, Owen Muir, and Michael Sikorav for reviewing the manuscript

Illustrations by Eleanor Aiken

Table of Contents

List of Tables

List of Figures

List of Screeners

SECTION I

Introduction

CHAPTER 1

What is "Difficult-to-Treat" Depression?

MOST CLINICIANS KNOW THE FRUSTRATION: A patient with depression starts an antidepressant, maybe two, and doesn't get better. You switch, you augment, and still—limited progress. You start to wonder if you're missing something. You are not alone.

The term "treatment-resistant depression" (TRD) is meant to describe these cases—specifically, when two adequate trials of antidepressants fail. But that term is both too broad and too narrow. It lumps together patients who need completely different approaches (like someone with bipolar spectrum features vs someone with vascular depression), and it leaves out patients who respond initially and then relapse, or who never had a clean starting point to define an "episode."

This book was born out of a need for a better way to think about these cases.

In 2002, a group of researchers gathered in San Francisco and coined a new term: Difficult-to-Treat Depression. They saw depression not as an acute illness to cure, but as a chronic condition to manage—more like diabetes than pneumonia. The goal was to improve function and quality of life, not to chase elusive symptom remission.

That's the spirit of this book. I've tried most of the therapies in it in my own practice—from lithium to light therapy, from pramipexole to probiotics. Some helped a lot, some only a little, but each has a place when you work with patients who are stuck.

What makes depression "difficult" is not just a lack of medication response. It's the diagnostic uncertainty, the comorbidities, the psychology of hopelessness, the loss of momentum, and the mismatch between the patient's life and the treatments we're offering.

This guide is organized to mirror the clinical approach. We begin with assessment and diagnostic challenges (Part I), then move through various treatment modalities, from psychosocial interventions to pharmacology, natural therapies, and neuromodulation (Part II). You won't find one algorithm that works for every case in these pages. What you will find is a new way of thinking—more flexible, more practical, and more collaborative. We'll take a hard look at the data—and at our own habits and expectations as clinicians.

If you've ever found yourself wondering *"Now what?"* when faced with a patient who isn't getting better, this book is for you.

NOTE:
Although I prefer the term *difficult-to-treat depression,* I occasionally use the more specific term *treatment-resistant depression* when referring to patients who did not have a meaningful response to two or more antidepressant trials.

TABLE 1-1. Common Features of Difficult-to-Treat Depression

Long duration of illness	Medical and psychiatric comorbidities
Frequent recurrence	Suicide risk
Incomplete recovery	Soft bipolar features
Multiple treatment failures	Early childhood adversity
Periods of disability and hospitalizations	

CHAPTER 2

A Primer on "Regular" Depression

BEFORE WE DIVE into difficult-to-treat depression, let's make sure we're not skipping the basics. Many cases that end up with that dreaded label are simply undertreated—poor dosing, rushed switches, no measurement, or just a failure to look under the hood. This chapter walks through a practical, real-world approach to treating "regular" depression—starting with a good evaluation and ending with a thoughtful treatment plan.

The Evaluation

The workup for depression begins before a prescription is written. Depression is not one illness—it's a syndrome. To treat it effectively, we need to ask: *Why is this person depressed?*

A thorough initial evaluation includes:

Timeline

Clarify the onset, duration, and course of the current episode. Ask about prior episodes, past treatments, and whether the patient recovered fully in the past or remained symptomatic between episodes.

Syndrome differentiation

- Do they meet DSM-5 criteria for major depressive disorder?
- Check for bipolar disorder, psychiatric comorbidities, and medical mimics (more on those in the next chapter).

Symptom profile

Look for atypical features (hypersomnia, weight gain, rejection sensitivity), melancholic features (early morning awakening, low appetite, psychomotor slowing or agitation), and comorbid symptoms like anxiety, cognitive impairment, and irritability.

Psychosocial context

Assess recent stressors, trauma history, interpersonal conflict, and protective factors.

The Treatment Plan

Once you've confirmed a diagnosis, the next step is a treatment plan—not just a prescription.

Set expectations

Many patients (and clinicians) underestimate how long it takes for treatment to work. Antidepressants usually take 2–6 weeks for initial effects and up to 12 weeks for full remission (APA Guidelines for Major Depressive Disorder, 2010). Partial improvement is common in early phases. Make sure your patients understand that trajectory.

Decide on medication

Start medication if:

- The depression is moderate to severe
- There's functional impairment
- Suicidality is present
- Therapy alone has failed
- The patient prefers medication

For medication-naïve patients with uncomplicated depression, the ideal agent is effective, well tolerated, and free of sexual side effects, weight gain, and sedation. **Bupropion (Wellbutrin)** often checks all those boxes—and it improves cognition (Gualtieri CT and Johnson LG, *MedGenMed* 2007;9(1):22). While it's not approved for anxiety disorders, it's effective for the nonspecific anxiety that often accompanies depression. The downsides include initiation insomnia (although it deepens sleep quality), and at doses above 300 mg/day, a small seizure risk.

Many clinicians are accustomed to starting with a selective serotonin reuptake inhibitor (SSRI), in part because they are highly effective for anxiety disorders and because we have long experience with their use. **Escitalopram (Lexapro)** and **sertraline (Zoloft)** are often first-line choices—both have minimal side effects and low risk of drug interactions.

See Table 2.1 for a list of clinical scenarios that often sway clinicians to choose a particular antidepressant, based on the clinical picture and the potential side effects of the medication.

TABLE 2-1. Choosing an Initial Antidepressant Based on Clinical Features

Clinical Scenario	First-Line Recommendations
Fatigue	Bupropion
Comorbid anxiety disorder	Escitalopram, sertraline
Insomnia, weight loss	Mirtazapine
Cognitive problems	Bupropion, vortioxetine
Sexual side effects are a dealbreaker	Bupropion, mirtazapine, or vortioxetine
Weight gain is a dealbreaker	Bupropion or fluoxetine
Chronic pain, fibromyalgia	Duloxetine
Smoking cessation	Bupropion
ADHD + depression	Bupropion
Bulimia	Fluoxetine
Atypical features (hypersomnia, etc.)	Monoamine oxidase inhibitors (MAOIs, often underused, but effective)
Child or adolescent	Fluoxetine

Titrate to a therapeutic dose, but don't go too high. As we'll see in Chapter 15, most antidepressants do not work better beyond the mid-range dose. Stay the course for 6–12 weeks unless intolerable side effects arise.

Psychotherapy: When, Why, and Which One

When therapy alone may be enough

Psychotherapy is reasonable as monotherapy for patients with mild to moderate depression. Predictors of a good psychotherapy response include secure attachment style, strong social supports, intact cognition, high distress, and a high motivation for therapy.

When therapy must be added

In cases of partial response to meds, comorbid personality disorders, chronic or recurrent depression, or life stress that overwhelms coping, therapy is essential. Therapy plus medication outperforms either alone in chronic depression and in moderate to severe depression (Cuijpers P et al, *World Psychiatry* 2020;19(1):92–107).

Which therapy?

Cognitive behavioral therapy (CBT) is often upheld as the evidence-based therapy for depression, but other approaches are just as effective. Interpersonal psychotherapy (IPT) is an eclectic approach that was developed to mimic what the average therapist does in practice: problem-solving conflicts in relationships and processing grief. Psychodynamic therapy, mindfulness-based therapy, and behavioral activation also treat depression.

Therapy works best when it draws on the patient's strengths. For example, patients who lack relationships don't do as well with IPT, and those with difficulties in logical thought don't do as well with CBT (Sotsky SM et al, *Am J Psychiatry* 1991;148(8):997–1008).

Lifestyle Interventions

Lifestyle changes aren't optional add-ons. They're evidence-based components of depression treatment. In animal studies, antidepressants do not work if the animal is kept in isolation, unable to socialize, or if the researchers remove the hamster wheel so it can't exercise. In Chapter 14 we'll look at eight lifestyle changes that augment antidepressants.

Is This Depression Really Difficult to Treat?

Use a checklist approach:

- Were two or more antidepressants tried at adequate dose and duration (6–12 weeks)?
- Were they taken consistently?
- Were comorbidities addressed?
- Were lifestyle factors optimized?
- Was the diagnosis correct?

If any of these are "no," you're not dealing with TRD—you're dealing with a case that hasn't had a full shot at recovery.

Case Vignette: Not-so-Difficult Depression

Demetrius, a 38-year-old nurse, was referred for TRD after he did not respond to fluoxetine and venlafaxine. Chart review showed fluoxetine was stopped at 10 mg after two weeks due to fatigue, and venlafaxine had only reached 75 mg. During treatment, he drank two to three glasses of wine nightly, had insomnia, and no rating scales were ever used. We restarted fluoxetine and titrated to 20 mg while initiating CBT and limiting alcohol. His depression rating scale (PHQ-9) dropped from 17 to 5 over 8 weeks.

Demetrius didn't have TRD. He had *undertreated depression.*

Takeaway

Before escalating to augmentation or neuromodulation, take the time to build a solid treatment foundation. That means a proper diagnosis, a deliberate plan, and full-dose, full-length treatment with measured outcomes.

SECTION II

Assessment and Diagnosis

Common Causes of Treatment Resistance

CHAPTER 3

Misdiagnosis

LET'S FACE IT—SOMETIMES THE FAILURE isn't the antidepressant but our own diagnostic prowess. When a patient isn't responding, consider these four common culprits:

- Missing bipolar disorder
- Missing psychiatric comorbidities
- Missing medical comorbidities
- Misjudging antidepressant response

Bipolar Disorder: The Great Pretender

Nearly half of patients with supposed "treatment-resistant depression" actually have undiagnosed bipolar disorder. The average patient with bipolar disorder waits seven years for the correct diagnosis, cycling through multiple failed antidepressant trials along the way.

Look for these clues:

- Early onset depression (before age 20)
- Family history of bipolar disorder
- History of agitation, irritability, or mood worsening on an antidepressant

We'll dive deeper into this vexing diagnosis in Chapter 5.

Psychiatric Comorbidities: Depression Rarely Travels Alone

Mental illness often hides from those best suited to help it. It doesn't exercise good judgment—if you don't ask, it won't speak up. This is especially true when secrecy and shame are part of the disorder, as with obsessive-compulsive disorder (OCD), social anxiety, eating disorders, substance use, and trauma.

The low mood may actually be a state of demoralization after years of untreated:

- ADHD (causing chronic underachievement)
- Anxiety disorders (leading to avoidance and a limited life)
- Personality disorders (creating relationship turmoil)
- Addiction (with its cascade of social and medical consequences)

Some comorbidities respond better to specific medications:

- **OCD**—Clomipramine (Anafranil) or high-dose selective serotonin reuptake inhibitors (SSRIs)—typically twice the dose needed for depression alone).
- **Anxiety disorders**—Medium to high doses of SSRIs or serotonin-norepinephrine reuptake inhibitor (SNRIs)—aim for the upper range of the optimal depression dose (see Chapter 15 for dosing tables).
- **ADHD**—Bupropion (Wellbutrin) or viloxazine (marketed as Qelbree for ADHD in the US but approved for depression elsewhere).
- **Bulimia and binge eating**—Fluoxetine (Prozac, FDA-approved for bulimia at 60 mg/day), sertraline (Zoloft, 50–100 mg), or duloxetine (Cymbalta, 60–120 mg). Avoid bupropion due to increased seizure risk in bulimia.
- **Substance use disorders**—Often require specialized treatment alongside depression care. Even moderate alcohol use can sabotage antidepressant response. Cannabis use correlates with poorer outcomes in depression treatment, particularly with daily use (Nunes EV et al, *Am J Psychiatry* 2023;180(3):179–181; Bahorik AL et al, *J Affect Disord* 2017;213:168–171). Consider asking:
  - "Many people use substances to cope with depression. What have you found helpful?"
  - "When did you last go for at least a month without alcohol or cannabis?"
  - "Have you noticed differences in your mood during periods of sobriety?"

Medical Disorders: The Body-Mind Connection

Medical disorders rarely cause depression on their own but frequently contribute to symptoms and treatment resistance. Consider these common culprits:

- **Sleep apnea**—Present in up to 40% of treatment-resistant cases. Ask about snoring, gasping for air, unrefreshing sleep, and daytime fatigue. Consider a sleep study for patients with risk factors (obesity, thick neck circumference, and older age; also, comorbid post-traumatic stress disorder [PTSD]).
- **Metabolic disorders**—Insulin resistance affects the brain, where insulin receptors are involved in cognition and mood regulation. Type 2 diabetes and pre-diabetes further drive depression by creating an inflammatory state.
- **Chronic pain**—Creates a "depression amplifier" effect, making symptoms feel worse and treatments less effective. Pain and depression share common neurobiological pathways.
- **Neurological disorders**—Small vessel cerebrovascular disease, early Parkinson's, and dementia can all manifest initially as depression.
- **Endocrine disorders**—Thyroid dysfunction, even subclinical, can mimic or exacerbate depression.
- **Medication effects**—Depression is a common side effect of corticosteroids, interferon, isotretinoin (Accutane), and progestin-containing contraceptives.

The first step is educating patients that physical health problems contribute to depression and can prevent antidepressants from working. Collaborate with their primary care provider. I often send a brief note explaining how their medical condition might be affecting their psychiatric treatment.

Laboratory Panel: Basic Workup

Consider checking these in patients who are older, in poor health, or who haven't responded well to antidepressants:

- Thyroid (TSH)
- Liver function (AST, ALT)
- Electrolytes
- Complete blood count
- Ferritin (more sensitive than hemoglobin for iron deficiency; aim for > 45 ng/mL, higher than the cutoff for anemia) (Stewart R et al, *Psychosom Med* 2012;74(2):208–213)
- Folate and B12
- Vitamin D (25-hydroxy)

- High-sensitivity C-reactive protein (hs-CRP, a marker of inflammation)
- Fasting glucose or HbA1c
- In men: Total testosterone level (especially if low energy, reduced libido)
- In women: Consider menopausal status (hormone levels usually unnecessary)

When abnormalities are found, address them systematically. Sometimes the improvement in depression is dramatic—such as when treating hypothyroidism or low folate. More often, it's incremental, but these small gains add up.

Misjudging Antidepressant Response: The Memory Problem

Depression distorts memory in a negatively biased way. Patients often forget periods of wellness or attribute improvements to external factors rather than medications.

When options seem exhausted, reassess the past medication list. Ask relatives about past response, check old records, and look for markers of improvement. Did the medication help them return to work or stay out of the hospital for a long time?

In the next chapter, we'll explore how to use assessment tools to overcome these memory problems and improve your treatment success rates.

Key Takeaways

- Before jumping into strategies for treatment resistance, check for bipolar disorder and psychiatric and medical comorbidities.
- Patients often forget what worked in the past, so dig deeper for signs of response.

CHAPTER 4

Optimal Use of Assessment Tools

SELECTING THE RIGHT TOOLS TO MEASURE progress separates the clinicians who achieve remission from those who settle for partial response. Let's explore which assessment approaches are worth your time and which aren't.

Pharmacogenetic Tests: Not Ready for Prime Time

Pharmacogenetic tests look at two types of genes:

- Pharmacokinetic genes, which code for enzymes that metabolize drugs. They tell us whether a person might have unusually high or low levels at normal dosages of a medication.
- Pharmacodynamic genes, which predict how the body and brain will respond to a medication once blood levels are in the therapeutic range.

These tests seemed promising in early trials, but those studies did not blind patients to the use of the test. When proper blinding was implemented in later research, the benefits evaporated. As of 2025, pharmacogenetic panels have failed to improve outcomes in multiple large, double-blind, randomized controlled trials for depression—even when enrolling patients who are most likely to benefit: those who did not recover after their first antidepressant trial (Baum M et al, *Am J Psychiatry* 2024;181(7):591–607).

Why have these tests failed? Let's start with the pharmacokinetic genes. These give only a rough approximation of drug metabolism because genes do not fully control enzyme activity. Age, gender, diet, smoking, and other medications account for 25%–40% of enzymatic function (Aiken C, *Med-Central*, Jun 6, 2023). Furthermore, estimating the metabolism is not very helpful because most antidepressants don't have strong dose-response relationships. If they did, we'd check serum levels rather than genes (as we do for tricyclics).

As for pharmacodynamic genes like the serotonin transporter (SERT), which supposedly predicts selective serotonin reuptake inhibitor (SSRI) response, results are inconsistent and meta-analyses show no reliable effect. The American Psychiatric Association (APA) does not recommend pharmacogenetic testing in depression. If you suspect unusual drug metabolism, check drug levels directly. It's cheaper and more accurate than genetic testing.

Measurement-Based Care: Double Your Remission Rates

While genetic testing disappoints, a simple paper-and-pencil approach can dramatically improve outcomes. In a 6-month study of 120 patients randomized to treatment with or without routine measurement, remission rates more than doubled among those who rated their depression at each visit (29% vs 74%). Both groups used the same medications (paroxetine [Paxil] or mirtazapine [Remeron]), but the measurement group followed an algorithm that adjusted the treatment based on self-reported scales, while the control group relied on clinical impression alone (Guo T et al, *Am J Psychiatry* 2015;172(10):1004–1013).

Why Measurement Works

Depression impairs the ability to detect change. We see this dulling even at the sensory level—depressed patients have trouble distinguishing different shades of gray in visual perception studies (Bubl E et al, *Biol Psychiatry* 2010;68(2):205–208). Add memory deficits and negative cognitive bias, and you understand why it's not enough to ask, "Do you feel better on the new medication?"

This holds true for the past medication history as well. I have seen many patients recover after retrying a medication that they recalled no benefit from. Sometimes relatives have a clearer recall for those lost gems. When options seem exhausted, bring in the relatives and reassess the past medication list.

Four Perspectives on Treatment Response

There are four ways to assess medication response:

1. **Patient's subjective report:** Does the patient think the medication is helping?
2. **Relative's observations:** Ask them directly or ask the patient if others have noticed changes.

3. **Your clinical impression:** This is your overall sense. It is informed by their mental status and functional improvements. Are they doing more? Less bothered by daily hassles?
4. **Rating scales:** Have patients complete the same scale before each visit, regardless of whether their treatment changed. Ratings at routine visits help create a baseline from which to measure change.

It is rare that these four viewpoints line up in the same direction. Even the patient's subjective report often differs from the changes on their self-rated scales. No single viewpoint holds the truth, and each needs to be assessed on its own terms. Over time, patterns emerge.

Implementing Measurement-Based Care

You don't need complex tools. Many clinicians use depression-focused scales like the PHQ-9 or its shorter version (PHQ-2):

TABLE 4-1. Patient Health Questionnaire-2 (PHQ-2)

*Over the **last two weeks,** how often have you been bothered by the following problems?*	Not at all	Several days	More than half the days	Nearly every day
Little interest or pleasure in doing things	0	1	2	3
Feeling down, depressed, or hopeless	0	1	2	3

Interpretation

- Score 0–2: Minimal symptoms
- Score 3–6: Clinically significant symptoms

I prefer a broader scale that captures symptoms across multiple comorbidities, and I created this General Symptom Scale to assess that:

TABLE 4-2. General Symptom Scale

Over the ***past week*** have you had problems with	Not at all	Mild		Moderate		Severe	
Depression, low energy/motivation, or lack of pleasure?	0	1	2	3	4	5	6
Anxiety, nervousness, or excessive worry?	0	1	2	3	4	5	6
Irritable, agitated, angry, quick to argue?	0	1	2	3	4	5	6
Trouble falling or staying asleep?	0	1	2	3	4	5	6
Poor concentration, attention, or distractibility?	0	1	2	3	4	5	6

This scale offers several advantages. It captures symptoms like anxiety that matter greatly to patients but aren't part of the diagnostic criteria for depression. It also allows patients to complete the same scale regardless of their diagnosis, which eases its adoption in a busy practice.

Making Measurement Work in Practice

Collecting scales routinely is a logistical challenge. Most electronic medical records can send an electronic scale before the visit, and office staff can provide paper copies at check-in. If those steps fail, I will have the patient complete the scale during the visit. This ensures the measures are complete and communicates the importance of the work.

One way to pick up on patterns is by graphing results over time and reviewing them with the patient. Often, a visual representation reveals gradual improvements that felt imperceptible to the patient. Looking at the graph together engages patients in their recovery. It also serves as a kind of cognitive therapy, challenging depressive beliefs with the reality of their own ratings.

The challenge isn't easy. Patients may dismiss a clear pattern of improvement with comments like "I never know how to answer those rating scales . . . it's just random numbers." Responses like that remind us that negative cognitive bias is a real thing, and it makes patients miss opportunities in life.

When Assessment Sources Conflict

When your four perspectives don't align (and they rarely do), trust the trend more than any single measurement. Value functional improvements, like returning to work or socializing, over symptom reports. Consider that improvement in one domain may precede others—activity levels often improve before mood.

Beyond Standard Scales

In specific situations, you might consider more targeted assessments. For cognitive symptoms, THINC-it is a validated, at-home app that measures cognitive change in depression (developed with funding from Lundbeck pharmaceuticals). If dementia is a concern, the Montreal Cognitive Assessment (MoCA) is a useful screen. Sleep problems can be assessed with two weeks of sleep diaries or data from wearable sleep trackers. And functional capacity can be measured with the FAST (Functional Assessment Short

Test) or the WHO Disability Assessment Schedule (WHODAS 2.0, brief version).

In the next chapters, we'll look at how specific subtypes of depression can complicate diagnosis and lead to treatment resistance. These include bipolar depression (30%–40%), mixed features (25%), psychotic depression (20%), inflammation (30%), vascular depression (50% after age 65), and those with past trauma (60%). Those percentages reflect the rates in general depression. They are even higher in treatment-resistant cases. Each steers treatment in a different direction, and each is easy to miss.

Key Takeaways

- Depression impairs the ability to detect change, so you can't rely on patient's report. Relatives' observations, clinical impression, and rating scales add valuable information.
- Rate symptoms at every visit. When treatment is guided by these numeric trends, it can double remission rates.
- The APA does not recommend pharmacogenetic testing for depression, and neither does *The Carlat Report*. If you suspect unusual metabolism, check the drug level directly.

Personalized Treatment for Depressive Subtypes

CHAPTER 5

Bipolar Depression

UNRECOGNIZED BIPOLAR DISORDER is the most common reason for lack of response to antidepressants. Careful assessment finds it in 40%–60% of people with treatment-resistant depression (Francesca MM et al, *Clin Pract Epidemiol Ment Health* 2014;10:42–47). If that number seems high, it may be because the line between bipolar and unipolar has changed significantly over the years.

A Brief History of the Bipolar Diagnosis

Bipolar disorder was first described by Emil Kraepelin, who distinguished it from schizophrenia in 1913. Kraepelin lumped bipolar and recurrent depression together as *manic-depressive illness*. He tried to divide them further, but he was unable to. "This classification, apparently so simple, really encounters manifold difficulties," he wrote, concluding that mania and depression were just different faces of the same illness.

Enter Jules Angst, born just two months after Kraepelin's death in 1926 (and still alive at the time of this writing). Angst believed that patients with mania differed from those who only had depression. They had different family histories, course of illness, and treatment response. In 1966, he proposed separating the mood disorders based on manic symptoms, and this bipolar-unipolar split was formally adopted in DSM-III in 1980. Before that, the manual had lumped bipolar and unipolar together under the single diagnosis of manic-depression.

Criticisms of the split appeared within months of DSM-III's publication. Psychiatrists argued that there was too much overlap in the genetics, treatment response, and even EEGs to warrant two categories. They called for a spectrum approach, and leading the charge was an unlikely physician: Jules Angst.

Over the next 30 years, Angst focused on the overlap between bipolar and unipolar disorders. His work was adopted in DSM, as disorders that bridge the gap between bipolar I and major depression were gradually added. By 2013, most of the spectrum disorders were absorbed into DSM-5, and the manual's chair, David Kupfer, described depression and bipolar as part of "a continuum, with variable expressions of vulnerability to hypomania or mania." (Phillips ML and Kupfer DJ, *Lancet* 2013;381(9878):1663–1671).

DSM-5 includes six categories that—when laid out in a line—form a spectrum of mood disorders. On one side are people with full mania (bipolar I), and on the other side are those with depression who never have manic or hypomanic symptoms.

Bipolar Disorders			Unipolar Depression		
Bipolar 1	Bipolar 2	Uni-polar	Brief hypomania	Mixed Features	Pure depression (no hypomania)

FIGURE 5-1. DSM Spectrum of Mood Disorders

In this way, we've returned to Kraepelin's original idea, where mood disorders are viewed as a single illness. From that perspective, it makes little sense to argue about whether a patient has bipolar or unipolar. Instead, we'd ask, "How bipolar are they?" The answer is not black and white, but it helps predict how they will respond to antidepressants.

A Spectrum Approach to Treatment

In theory, the bipolar diagnosis is supposed to tell us whether a patient will respond to mood stabilizers or antidepressants. In practice, it is less clear. Some patients with bipolar disorder improve on antidepressants, and some with unipolar depression get worse on them. These patterns make more sense if we adopt a spectrum approach, as a study by Jules Angst and colleagues shows.

In Figure 5.2, the risk of mood worsening on an antidepressant rises gradually with the degree of hypomania. At the right are patients with DSM-5 bipolar disorder, which requires at least four days of hypomania or seven days of mania to make the diagnosis. Here, 30%–40% developed manic or hypomanic switches on an antidepressant. Those with subthreshold bipolarity (hypomanias lasting one to three days), also had antidepressant-induced

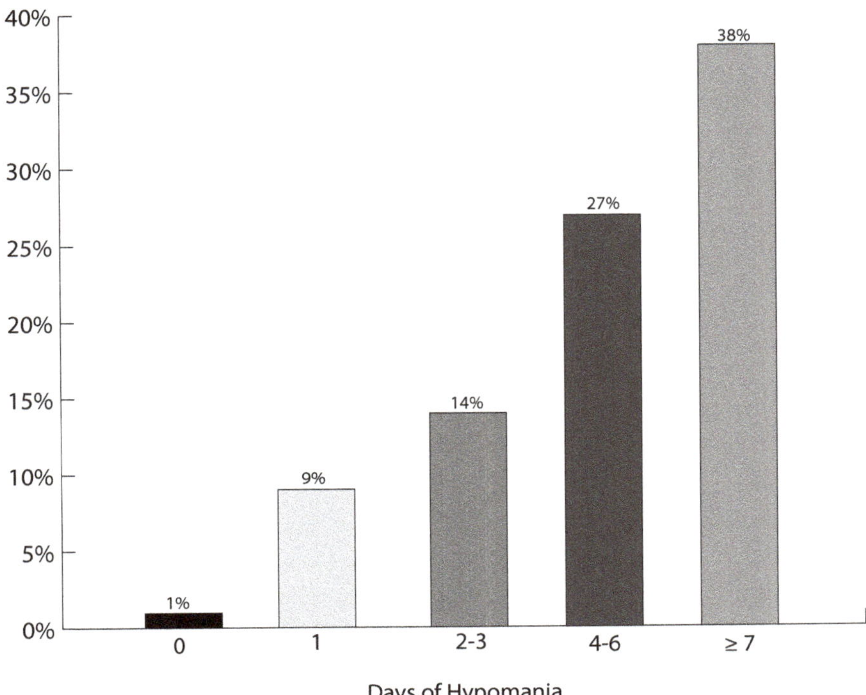

FIGURE 5-2. Risk of antidepressant-induced mania/hypomania rises with duration of past manic/hypomanic symptoms

switches, but not as often. If we graphed other markers of bipolarity, such as family history, we would probably see similar results.

The more bipolar features a patient has, the more likely their mood will worsen on an antidepressant and improve on a mood stabilizer. Supporting that idea is the work of Gary Sachs, whose Bipolarity Index scale tallies how many bipolar features a patient has. The higher the score, the more likely they are to respond to a mood stabilizer (Del Debbio A et al, *Bipolar Disord* 2007;9:S32–S33).

Cyclothymic Disorder

In the middle of the spectrum is cyclothymic disorder. These patients experience frequent, brief mood swings from sluggish to energized, fearful to impulsive, extraverted to withdrawn. That inconsistency makes them feel insecure, unsure of how they'll be from one day to the next. As one patient put it, "I could do well on any job for a few months, but then I'd fall apart." Their interests are broad, but they can become a "Jack or Jill of all trades but master of none." The brighter sides of this temperament may draw people into their circle, but conflict seeps in as others find out they are not dependable.

Although cyclothymia has been recognized since at least 1883, it is often missed in practice. Its many faces can also lead to diagnostic confusion. For example, the impulsive, restless, and distracted nature of cyclothymia can resemble ADHD, or it may be mistaken for an anxiety disorder. Cyclothymia is also a common cause of substance abuse.

One reason for its under-recognition is that DSM-5 views it as a separate, categorical disorder and does not allow it to be diagnosed in patients with full bipolar disorder. Most research, however, views cyclothymia as a temperament that underlies other bipolar disorders, particularly bipolar II, where 38% of patients have this temperament. Patients with both bipolar II and cyclothymic temperament have a stormier course, with more impulsivity, anxiety, self-harm, and unstable relationships (Akiskal HS et al, J Affect Disord 2003;73(1–2):49–57). The clinical picture resembles borderline personality disorder (BPD), something noted by researchers from Hagop Akiskal to Marsha Linehan (Aiken C, *Psych Times*, May 16, 2019).

Three Steps to a Bipolar Diagnosis

Whether you take the categorical or the spectrum view, bipolar disorder is a difficult diagnosis to make. On average, patients wait seven years before getting an accurate diagnosis. The FDA recommends screening for bipolar disorder before prescribing an antidepressant, and that's a good place to start.

1. Screening instruments

Popular screening instruments for bipolar disorder include the Mood Disorder Questionnaire and the Bipolar Spectrum Diagnostic Scale, both of which are readily available online. These tests have high specificity, which means when they are positive, they are usually right. But the problem is they miss too many cases. We need a screener that casts a wide net.

Screening is just the first step in diagnosis, and we can rule out any false positives at the next step.

The Rapid Mood Screener (RMS) is a newer test that accomplishes that (Figure 3.2). While most screeners miss 30% of bipolar cases, this one only misses 10% (ie, its sensitivity is 90%). It also has a respectable specificity of 80%, which means that when it is positive, it is right 80% of the time (McIntyre RS et al, *Curr Med Res Opin* 2021;37(1):135–144). What makes this screener different is that it looks for both manic symptoms and nonsymptomatic markers of the illness, like age of onset and antidepressant response.

The RMS has been validated in both bipolar I and II, but it was developed for bipolar I disorder and therefore requires at least a week of manic symptoms (Liao Y et al, *J Affect Disord* 2024;348:54–61). I have modified the scale to capture a broader spectrum of manic symptoms, and this version is printed here as the RMS-II.

SCREENER 5-1. Rapid Mood Screener II

Consider your whole life as you answer these questions		
Have there been at least six different periods of time (at least two weeks) when you felt deeply depressed?	Yes	No
Did you have problems with depression before the age of 18?	Yes	No
Have you ever had to stop or change your antidepressant because it made you highly irritable or hyper?	Yes	No
Have you ever had a period of time during which you were more talkative than normal with thoughts racing in your head? *If yes, what was the longest it lasted for?*		
No • Yes, 1–3 days • Yes, 4 or more days • Yes, 7 or more days		
Have you ever had a period of time during which you felt any of the following: unusually happy; unusually outgoing; or unusually energetic? *If yes, what was the longest it lasted for?*		
No • Yes, 1–3 days • Yes, 4 or more days • Yes, 7 or more days		
Have you ever had a period of time during which you needed much less sleep than usual? *If yes, what was the longest it lasted for?*		
No • Yes, 1–3 days • Yes, 4 or more days • Yes, 7 or more days		

Scoring: Positive if 4 or more YES's

The Bipolar Spectrum Diagnostic Scale (BSDS) is provided in Screener 5.2. Like the Mood Disorder Questionnaire (MDQ), it is symptom-focused, but it has better sensitivity for the hypomanias that characterize bipolar II disorder, the more common of the two types.

SCREENER 5-2. Bipolar Spectrum Diagnostic Scale

Instructions: Please read through the entire passage below before filling in any blanks.
Some individuals notice that their mood and/or energy levels shift drastically from time to time____. These individuals notice that, at times, their mood and/or energy level is very low, and at other times, very high ____. During their "low" phases, these individuals often feel a lack of energy; a need to stay in bed or get extra sleep; and little or no motivation to do things they need to do____. They often put on weight during these periods ____. During their low phases, these individuals often feel "blue," sad all the time, or depressed____. Sometimes, during these low phases, they feel hopeless or even suicidal____. Their ability to function at work or socially is impaired____. Typically, these low phases last for a few weeks, but sometimes they last only a few days____.
Individuals with this type of pattern may experience a period of "normal" mood in between mood swings, during which their mood and energy level feels "right" and their ability to function is not disturbed____. They may then notice a marked shift or "switch" in the way they feel____. Their energy increases above what is normal for them, and they often get many things done they would not ordinarily be able to do____. Sometimes, during these "high" periods, these individuals feel as if they have too much energy or feel "hyper" ____. Some individuals, during these high periods, may feel irritable, "on edge," or aggressive____. Some individuals, during these high periods, take on too many activities at once____. During these high periods, some individuals may spend money in ways that cause them trouble____. They may be more talkative, outgoing, or sexual during these periods____. Sometimes, their behavior during these high periods seems strange or annoying to others____. Sometimes, these individuals get into difficulty with co-workers or the police during these high periods____. Sometimes, they increase their alcohol or nonprescription drug use during these high periods____.
Now that you have read this passage, please check one of the following four boxes (consider your whole life when you answer, including recent times): ☐ This story fits me very well, or almost perfectly (6). ☐ This story fits me fairly well (4). ☐ This story fits me to some degree, but not in most respects (2). ☐ This story does not really describe me at all (0). Now please go back and put a check after each sentence that definitely describes your life.

Scoring

Add one point for each box checked in the paragraph, and 6, 4, 2, or 0 points for the final items. Scores ≥ 13 are suggestive of bipolar disorder.

Source: Ghaemi SN et al, *J Affect Disord.* 2005;84(2–3):273–277.

2. Structured interviews

A positive screen for bipolar disorder requires confirmation with a diagnostic interview. The gold standard is a structured interview like the Structured

Clinical Interview for DSM Disorders (SCID) or the Mini International Neuropsychiatric Interview (MINI). These may seem daunting and time-consuming, but they are not. They simply translate the DSM criteria into plain English.

Structured interviews are copyrighted. If you don't have access to them, you can translate the DSM criteria into YES or NO questions on your own. I've done that here in Table 5.1. This version includes questions that capture the mixed form of manic symptoms, such as the wired, anxious energy that makes people feel driven to do something but unsure of what to do.

TABLE 5-1. Aiken's Structured Interview for Mania and Hypomania

DSM Symptom	Screening Question
Elevated mood	"Have you ever felt more up, driven, and/or* energized than usual, even if it was an uncomfortable, wired, anxious energy?"
Irritability	"Have you ever been angry, irritable, or quick to argue for several days?"
If they endorse *elevated mood* or *irritability*, ask about the associated symptoms below. Look for stretches of time when they were more or less this way, most of the time. Manic symptoms are in rapid flux, so the symptoms may not overlap 100%. They may occur during depression in a mixed state. "When you felt [elevated mood/energy/irritability—use their words] did you also . . ."	
Confidence	". . . feel more confident, like you can achieve great things? Or more demanding, pushy, sure of yourself?"
Short sleep	". . . sleep less than usual but still kept going the next day, even if you wished you could get more sleep?"
Talkative	". . . talk more than usual? Did people think you were loud or difficult to interrupt?"
Racing thoughts	". . . have racing thoughts, multiple trains of thought, or thoughts that jumped around so quickly that it was hard to keep up with them?"
Distracted	". . . get easily distracted by little things around you or your own thoughts? Jump quickly from one activity to another?"
Hyperactive	". . . do a lot more than usual, like taking on more projects or being more socially or sexually active? Feel more driven or restless even if you didn't know what to do with the drive?"
Impulsive	". . . jump into things without thinking about the consequences? Do things that others thought were extreme, impulsive, or out of character for you?" [List examples "overspending, aggressive driving, sudden travel, breaking the law or breaking things, or doing things that could hurt your job, school, relationships, or your life]"
Duration	"Were you ever more or less this way [summarize the symptoms in their words] for four days in a row? For seven days in a row?"

TABLE 5-1. Aiken's Structured Interview for Mania and Hypomania

DSM Symptom	Screening Question
Consequences	"What was the biggest problem that these symptoms caused? What is the most extreme thing you did?" If they seem hesitant, remind them there is no shame here and suggest extreme examples like prison, bankruptcy, loss of jobs or relationships. **Hypomania** = mild-moderate problems. They were able to overcome the consequences without great difficulty (eg, traffic tickets, arguments, warnings at work, overspending within budget). **Mania** = severe problems. The consequences were potentially or actually irreversible or difficult to overcome. They could have led to arrest or hospitalization (eg, violence; major debt; loss of job, school, or relationships).
Scoring This interview is positive for bipolar disorder if they endorse: 1) Elevated mood and at least 3 out of 7 associated symptoms *or* 2) Irritable mood and at least 4 out of 7 associated symptoms The type of bipolar depends on the duration and severity: *Short duration hypomania:* longest duration less than 4 days and problems were mild-moderate (if the hypomanic symptoms cycle with brief depressions it is cyclothymic disorder). *Hypomania:* longest duration at least 4 days and problems were mild-moderate. Diagnosis is bipolar II if they also had a history of full depression. *Mania:* longest duration at least 7 days and consequences were severe or psychotic symptoms were involved. Duration can be less than 7 days if the symptoms led to hospitalization. Diagnosis is Bipolar I if any mania, regardless of whether there was depressive episode.	

3. Family input

People with mania live in the moment, and they often forget those moments by the time they present with depression. Not so for the family, who recall it with a clearer eye, particularly if the manic state was traumatic for them. When I ask about manic symptoms, I invite the family to answer as well. If they can't come in, I'll dial them up on telehealth or have them complete a self-rated scale for the patient.

Diagnostic Dilemmas

As you gather all this information, you'll often find conflicting results. Family and patients disagree, or patients say "yes" to a question on paper that they then say "no" to in person. Symptoms of mania and hypomania are hard to pin down. They overlap with other disorders ("distracted," "racing thoughts") and with normality (hopefully everyone has had times of high energy and motivation). Here are some tips to guide you.

Focus on Energy, Not Emotions

Energy is the core symptom that divides mania (high energy) from depression (low energy). Emotions, in contrast, vary widely and are poorly recalled. In the depths of depression, it's difficult to recall a time of euphoria. Even in a mania, euphoria is rare. More common are rapid emotional swings (lability) and a frightening sense that one has lost control. That does not feel good, which may explain the surprising finding that anxiety is more common during mania than during depression. As psychologist Kay Redfield Jamison put it, "I felt infinitely worse [during mania] than when in the midst of my worst depressions." Likewise, when patients who were hospitalized for mania were asked why they were there, the most common answer was "depression" (Kotin J and Goodwin FK, *Am J Psychiatry* 1972;129(6):679-686).

Probe Ambiguous Answers

Structured interviews call for "yes" or "no" answers, so how do we interpret responses like "sort of," or "not really." Well, if you asked someone if they ever stole from their friends, and they answered, "Not really," you'd take that as a serious sign. All the more so when the choice of a brain-active medication hinges on the answer. Here's how to interpret these uncertain areas.

TABLE 5-2. Structured Bipolar Interview: Ambiguous Answers

Ambiguous Answer	Interpretation
"Not in a long time" "Very rarely" "Almost never"	These answers are consistent with bipolar. People with bipolar spend only 4%–15% of their lifespan in manic or hypomanic states.
"Yes, but that's just the way I always am." "Yes, that's how I am when I'm not depressed. I have energy and motivation."	These answers make bipolar less likely, but do not rule it out. Hypomania reflects a change from the person's usual state, but many patients have affective temperaments, meaning their baseline personality is a little hypomanic or irritable.

TABLE 5-2. Structured Bipolar Interview: Ambiguous Answers

Ambiguous Answer	Interpretation
"Yes, but only when I'm with a lot of friends, having a good time, or when I have a lot to do."	This means "Yes." Avoid downgrading symptoms because they have an understandable explanation. People are more likely to socialize and take on projects when hypomanic. Likewise, when someone has felt depressed for two weeks "but only because I lost my job," we still code it as depression.
"Yes, I do impulsive things [sex, drugs, travel, or retail therapy], but that's because it's the only way I can feel better when depressed"	This means "Yes." As with the above, beware of contextualization. Hedonic behaviors are rare during depression, and their appearance is suggestive of a mixed state.
"Yes, I'm often distracted with racing thoughts. That's my ADHD."	This means "Yes." Comorbidities like ADHD and anxiety are common in bipolar disorder and should not be used to explain away the symptoms.

A Spectrum Approach to Diagnosis

Even the most careful workup often ends in uncertainty. Patients and relatives disagree. Comorbidities confuse the picture, and unambiguous answers are in short supply. The spectrum approach comes in handy here. First, it allows you to rank symptoms as clearly positive, clearly negative, or unclear. Second, it supports the symptomatic assessment with markers for the illness, such as age of onset, family history, and treatment response.

I'll review two ways to bring a spectrum approach to your diagnosis.

The Comprehensive Approach

The most comprehensive spectrum tool is the bipolarity index. This includes most of the cardinal signs and symptoms of bipolar disorder, divided into five sections:

- Episode characteristics
- Age of onset
- Course of illness
- Treatment response
- Family history

Each section is scored from 0–20 for a total score of 0–100 (only score the highest number in each section). A score of 50 or higher indicates a high probability of bipolar disorder with 90% sensitivity and specificity. The index is meant to be rated by the clinician, but I invite the patient in to review the work with me. Patients are less likely to reject the results when they are engaged in the process.

SCREENER 5-3. The Bipolarity Index

Directions: Circle the bulleted items that are positive in the patient's history. Score each of the five sections by circling the highest number (0–20) for which there is at least one positive item. The final score is the sum of all five sections.

I. EPISODE CHARACTERISTICS

Score	Criteria
20	• Acute manic episode with prominent euphoria, grandiosity, or expansiveness and no significant medical or other secondary etiology
15	• Acute mixed episode or dysphoric or irritable mania with no significant medical or other secondary etiology
10	• Hypomanic episode with no significant medical or other secondary etiology; or • Cyclothymic with no significant medical or other secondary etiology; or • A manic episode within 12 weeks of starting an antidepressant
5	• A hypomanic episode within 12 weeks of starting an antidepressant • Episodes with characteristic symptoms of hypomania, but symptoms, duration, or intensity are subthreshold for hypomania; or • A single MDE with psychotic or atypical features (atypical is ≥ 2 of the following: hypersomnia, hyperphagia, or leaden paralysis of limbs); or • Any postpartum depression
2	• Recurrent unipolar major depressive disorder (≥ 3 episodes); or • History of any kind of psychotic disorder (ie, presence of delusions, hallucinations, ideas of reference, or magical thinking)
0	• No history of significant mood elevation, recurrent depression, or psychosis

II. AGE OF ONSET (FIRST AFFECTIVE EPISODE OR SYNDROME)

Score	Criteria
20	• 15 to 19 years
15	• Before age 15 or between age 20 and 30
10	• 30 to 45 years
5	• After age 45
0	• No history of affective illness (no episodes, cyclothymia, dysthymia, or unspecified bipolar)

III. COURSE OF ILLNESS & ASSOCIATED FEATURES

Score	Criteria
20	• Recurrent, distinct manic episodes separated by at least 2 months of full recovery

SCREENER 5-3. The Bipolarity Index

15	• Recurrent, distinct manic episodes with incomplete inter-episode recovery; or • Recurrent, distinct hypomanic episodes with full inter-episode recovery
10	• Any substance use disorder (excluding nicotine/caffeine); or • Psychotic features only during acute mood episodes; or • Incarceration or repeated legal offenses related to manic behavior (eg, shoplifting, reckless driving, or bankruptcy)
5	• Recurrent unipolar major depressive disorder (MDD) with ≥ 3 major depressive episodes; or • Recurrent, distinct hypomanic episodes without full inter-episode recovery; or • Borderline personality disorder; anxiety disorder, including post-traumatic stress disorder (PTSD) and obsessive-compulsive disorder (OCD); eating disorder; or history of ADHD with onset before puberty; or • Engagement in gambling or other risky behaviors with the potential to pose a problem for patient, family, or friends; or • Behavioral evidence of premenstrual exacerbation of mood symptoms
2	• Baseline hyperthymic personality when not manic or depressed; or • Marriage 3 or more times (including remarriage to the same individual); or • In two or more years, has started a new job and changed jobs after less than a year; or • Has more than two advanced degrees
0	• None of the above.
IV. RESPONSE TO TREATMENT	
Score	**Criteria**
20	• Full recovery within 4 weeks of therapeutic treatment with a mood stabilizer
15	• Full recovery within 12 weeks of therapeutic treatment with a mood stabilizer or relapse within 12 weeks of discontinuing treatment; or • Affective switch to mania (pure or mixed) within 12 weeks of starting a new antidepressant or increasing dose
10	• Worsening dysphoria or mixed symptoms during antidepressant treatment subthreshold for mania (exclude worsening that is limited to known antidepressant side effects such as akathisia, anxiety, or sedation); or • Partial response to one or two mood stabilizers within 12 weeks of therapeutic treatment; or • Antidepressant-induced new or worsening rapid-cycling course
5	• Treatment resistance: lack of response to complete trials of three or more antidepressants; or • Affective switch to mania during antidepressant withdrawal
2	• Immediate, near-complete response to antidepressant withdrawal within one week or less
0	• None of the above, or no treatment

SCREENER 5-3. The Bipolarity Index

V. FAMILY HISTORY	
Score	**Criteria**
20	• At least one first-degree relative with clear bipolar disorder
15	• At least one second-degree relative with clear bipolar disorder; or • At least one first-degree relative with recurrent unipolar MDD and behavioral evidence suggesting bipolar disorder
10	• First-degree relative with recurrent unipolar MDD or schizoaffective disorder; or • Any relative with possible bipolar disorder or recurrent unipolar MDD and behavioral evidence suggesting bipolar disorder
5	• First-degree relative with clear substance use disorder (excluding nicotine/caffeine); or • Any relative with possible bipolar disorder
2	• First-degree relative with possible recurrent unipolar MDD; or • First-degree relative with anxiety disorder (including PTSD and OCD), eating disorder, or ADD/ADHD
0	• None of the above or no family history of psychiatric disorders

Total score (0 –100). Add the highest number in each section. A score ≥ 50 indicates a high probability of bipolar disorder.

The Rough-and-Ready Approach

If time is short, I'll use the Rapid Mood Screener or ask about three high-yield items from the Bipolarity Index:

"How old were you when you first experienced mood problems?"

Classic bipolar disorder begins between ages 15–20. However, when the patient has multiple psychiatric comorbidities, temperamental instability, or frequent mixed states, the onset is often in childhood or "as long as I can remember."

"Have you ever felt worse after starting an antidepressant?"

Here we want to rule out side effects. Any evidence of psychiatric worsening points to bipolar disorder. Overt switches into pure mania are rare. More often, the antidepressant sprinkles hypomanic symptoms on top of the depression, causing mixed features. For patients, this feels like a more anxious, agitated, dysphoric depression.

"Do any of your family members have bipolar disorder, or do they have depression along with anger problems, gambling, or do they

do impulsive things that have made you suspect they have bipolar disorder?"

Bipolar disorder is more heritable than major depression. The importance of this is illustrated by a study that tested paroxetine (Paxil) in children with MDD whose parent had bipolar disorder. The investigators halted the study midway because 75% of the children developed severe problems on the antidepressant, including mania, hospitalizations, and suicide attempts (Findling RL et al, *J Child Adolesc Psychopharmacol* 2008;18(6):615–621).

Case Vignette: The Spectrum in Practice

Melissa, a 32-year-old graphic designer, presented with her third episode of depression in five years. Her mood problems began at age 16. A previous trial of fluoxetine "kind of worked" but pooped out, while venlafaxine made her anxious and irritable.

When asked about potential hypomanic symptoms, she initially said "I've never been manic," but then described periods of heightened productivity between depressions when she would work 12-hour days, sleep 5 to 6 hours, and feel "finally normal."

Her husband, attending the second session, added that these "normal" periods concerned him because Melissa would become irritable when interrupted, take on too many projects, and spend excessively on art supplies she didn't need. These elevated episodes come on several times a year and usually last two to three days.

Diagnostic Assessment

- On the Rapid Mood Screener II, Melissa scored 4/6
- On structured interview, she met criteria for hypomanic symptoms but their duration was only two to three days
- Family history was positive for "moody depression" in her father and a cousin with diagnosed bipolar II disorder
- On the Bipolarity Index, her score was 45, just below the cutoff of 50

Spectrum-Based Treatment Approach

Rather than debating whether Melissa fits neatly into a diagnosis of bipolar or MDD, I explained the spectrum concept. We discussed how she fell into the "subthreshold bipolar" range with strong indicators for antidepressant-induced mood destabilization.

We agreed on a three-part plan:

1. Slowly taper venlafaxine to minimize withdrawal effects
2. Start lamotrigine, gradually titrating to a therapeutic dose
3. Implement regular wake times, an evening wind-down routine, and mood charting

At six-month follow-up, Melissa reported a more stable mood with less severe depressions and better control over her productive periods. She noted, "I still have my creative energy, but it doesn't take over my life anymore."

This case illustrates how recognizing subthreshold bipolarity can guide the treatment even when the patient doesn't meet full DSM criteria for bipolar disorder.

What to Do with the Antidepressant

If your patient has bipolar disorder, you can put this book down and move on to standard therapies for bipolar depression like lithium, lamotrigine (Lamictal), and certain second-generation antipsychotics (SGAs)—cariprazine (Vraylar), lumateperone (Caplyta), lurasidone (Latuda), olanzapine-fluoxetine combination (Symbyax), and quetiapine (Seroquel). Pramipexole (Mirapex) is a good option when you are on the fence about the diagnosis, as this dopamine agonist is effective in both bipolar and unipolar depression (see Chapter 21: Pramipexole).

That leaves a dilemma. Most patients are already taking an antidepressant when bipolar disorder is diagnosed. Antidepressants worsen the course of the illness, causing mixed features, rapid cycling, and—more rarely—overt mania. On the other hand, about 10%–20% of bipolar patients respond to antidepressants, something we see more often in bipolar II. However, the bigger risk in this scenario is often antidepressant withdrawal, particularly with serotonergic agents: selective serotonin reuptake inhibitors (SSRIs) and serotonin-norepinephrine reuptake inhibitor (SNRIs).

We don't have exact numbers on how common this problem is, but the longer the patient has been on the antidepressant, the more likely they are to have withdrawal problems. Antidepressant withdrawal looks a lot like mixed features, with intense anxiety, dysphoria, and lability. Paradoxically, abrupt withdrawal can trigger full mania.

The solution is first to figure out if the antidepressant is helping, harming, or doing nothing at all. Unless it is helping, taper off slowly, lowering by larger increments in the beginning, and smaller and smaller increments as you get near the end, following the guide in Chapter 46: Deprescribing Extra Antidepressants. How slowly will depend on the patient. Plan for at least three months with serotoninergic antidepressants, and at least one month with other classes.

When Mood Worsens on an Antidepressant

When patients call because their mood is worse after starting an antidepressant, don't assume they have bipolar disorder, and don't just tell them to soldier on. Here are other causes to consider.

Borderline Personality Disorder (BPD)

In BPD, antidepressants can help or harm. SSRIs improved mood, irritability, and impulsivity in small controlled trials of BPD. On the other hand, BPD is a risk factor for manic symptoms on antidepressants (Barbuti M et al, *J Affect Disord* 2017;219:187–192). When tricyclics were tested in BPD, some patients benefited, while others became more irritable, assaultive, disinhibited, paranoid, and suicidal (Soloff PH et al, *Psychopharmacol Bull* 1987;23(1):177–181).

Children, Adolescents, and Young Adults

In young patients, antidepressants raise the risk of suicidal ideation, though not of suicide attempts. The problem is rare (7–20 incidents per 1,000) and usually presents as pressured thoughts of suicide that come within weeks of starting an antidepressant. After age 25, the risk is no longer detectable, and after age 65 it reverses, and we see a protective effect with antidepressant therapy (Pompili M et al, *Pharmaceuticals* 2010;3(9):2861–2883; Sharma T et al, *BMJ* 2016;352:i65).

Undiagnosed bipolar disorder is one cause of this reaction, but most cases do not have mixed or manic symptoms. Other risk factors include younger age, female gender, past suicidality, the short arm (S) of the serotonin transporter (SERT), psychotic features, PTSD, substance use, and personality disorders (Sørensen JØ et al, *Acta Psychiatr Scand* 2022;145(2):209–222).

Tolerability and Slow Metabolizers

Antidepressant side effects can worsen mood, such as insomnia and akathisia on serotonergic antidepressants. Side effects can come on dramatically in patients who metabolize the medication slowly.

In rare cases, trazodone (Desyrel) can cause dysphoria, suicidality, and hallucinations due to a metabolite with MDMA-like properties (mCPP). This metabolite exits the body through CYP2D6, so the problem is more common when that enzyme is slowed, whether due to drug interactions (eg, fluoxetine [Prozac], paroxetine [Paxil], duloxetine [Cymbalta]) or genetic differences.

Course of Illness

Patients tend to start antidepressants because their mood is getting worse, and their mood might keep going in that direction if the medication doesn't help. This can make it look like the antidepressant is causing harm. In this case, soldiering on is the best advice, unless they've been at an effective dose for more than 3 to 6 weeks (or longer, up to 12 weeks, for a second antidepressant trial), in which case a change of treatment is in order.

Key Takeaways

- Around half of patients with "treatment-resistant depression" have unrecognized bipolar disorder.
- Assess for manic symptoms with a rating scale and structured interview.
- Look for nonsymptomatic signs of bipolar disorder like age of onset (15–20), family history, and mood worsening on antidepressants.

CHAPTER 6

Depression with Mixed Features

THE MIXED FEATURES SPECIFIER IS A NEW addition to DSM-5. It recognizes patients who have a few manic symptoms during their depression. Specifically, they have at least three manic symptoms, but not enough to qualify for a full diagnosis of mania or hypomania.

Mixed features occur in both bipolar and unipolar depression, and they are common in both, with rates of 24% in unipolar depression and 35% in bipolar depression (Vázquez GH et al, *J Affect Disord* 2018;225:756–760)*. This chapter will focus on the unipolar types.

Detecting mixed features requires a bit of finesse, as manic symptoms look different when they overlap with depression, much as mixing yellow and blue creates a completely new color (green).

Clinical Presentation: The Paradox of Mixed Features

How can someone feel euphoric, confident, and energized during depression? Let's look at those one by one:

- **Elevated mood**—In mixed depression, mood is often labile, swinging rapidly between irritable, sad, anxious, despairing, and—rarely—giddy or euphoric.
- **Elevated confidence**—Although depression has sunk their self-esteem, the patient's relatives may tell you about behaviors that betray a hidden confidence: demanding, intimidating, or stubbornly domineering in arguments.

---

*This study deviated from DSM-5 by allowing irritability, distractibility, and hyperactivity to count toward the diagnosis. DSM-5 made a controversial decision to exclude those manic symptoms from the mixed features count, reasoning that they are not specific enough to mania.

- **Elevated energy**—This is an uncomfortable, anxious energy that feels "wired, restless, crawling out of my skin."

Sleep is irregular during mixed states, vacillating between hypersomnia and hyposomnia. These patients don't identify with a "decreased need for sleep." They desperately want to sleep, to turn their mind off, which is crowded with depressive ruminations and worst-case scenarios.

While manic patients have a focused drive, the excess motivation in mixed features is directionless. "I feel driven to do something, but I don't know what to do." At times, it turns destructive, impulsively abandoning jobs, relationships, and breaking expensive electronics that a manic patient would have bought on credit. This dangerous combination of depression and impulsivity is one reason for the high risk of substance use and suicide in this population.

Assessment: Beyond Self-Report

During the interview, watch the patient as if they are in a silent movie. Do they fret and fidget, suggesting a high-energy state? Or are they passive and still? Now listen. With mixed features, their speech has a pressured, desperate, emotional tone, in contrast to the slow monotone of pure depression. How do they interact with you? Patients with pure depression are passive. Getting them to accept help is an uphill climb. In mixed features, there is a desperate drive for help. They call for an urgent appointment. They make demands for rapid relief and are difficult to reassure.

The Antidepressant Dilemma

In terms of family history, course of illness, and treatment response, mixed depressions fall somewhere between bipolar and unipolar. They are more likely to worsen on antidepressants, but not as much as a true bipolar patient, and some improve on them. With all of that uncertainty, antidepressants are second-line strategies in mixed depression, but most patients are already taking one when the mixed features are discovered. In those cases, the decision to taper or continue depends on their history:

- They clearly improved on the antidepressant. Keep it going.
- The antidepressant did not help, and the mixed features came on after starting it. Slowly taper off it.

- The antidepressant did nothing (or helped a little but wore off). Consider a slow taper, but it is best to wait until their mood improves with other treatment steps.
- The key is to taper slowly, particularly with serotonergic antidepressants, as withdrawal symptoms can mimic mixed features. See Chapter 46: Deprescribing Extra Antidepressants for more specific guidance on antidepressant tapering.

Treatment of Mixed Features

Mixed features predict a poor response to antidepressants, electroconvulsive therapy (ECT), and transcranial magnetic stimulation (TMS), but there are strategies that work (Tavares DF et al, *Neuropsychopharmacology* 2021;46(13):2257–2265; Perugi G et al, *Brain Stimul* 2012;5(1):18–24). Recovery here is difficult and slow, and these patients require more support along the way. Patients with mixed features tend to experience more medication side effects and less frustration tolerance. Often the manic symptoms improve before the depression, causing patients to give up prematurely unless we guide them on what to expect.

First-Line Pharmacotherapy

Second-generation antipsychotics (SGAs) have good evidence in depression with mixed features, particularly lumateperone (Caplyta) (42 mg/day), lurasidone (Latuda) (20–60 mg/day), and to a lesser extent ziprasidone (Geodon) (80–160 mg/day) and aripiprazole (Abilify) (2–10 mg/day). Most of these were tested as monotherapy, although they can also be added to an antidepressant (Durgam S et al, *J Clin Psychopharmacol* 2025;45(2):67–75; Goldberg JF et al, *CNS Spectr* 2017;22(2):220–227; Patkar A et al, *PLoS One* 2012;7(4):e34757; Han C et al, *Clin Psychopharmacol Neurosci* 2019;17(4):495–502).

Many experts rely on mood stabilizers for mixed depression, such as lithium, lamotrigine (Lamictal), valproate (Depakote), or carbamazepine (Equetro). Lithium is particularly helpful when suicidality is present. Lamotrigine is often well tolerated but may take 6–8 weeks for full effect. Valproate has a faster onset but more side effects. Carbamazepine has problematic drug interactions, but this anticonvulsant is reasonably well tolerated and has evidence to prevent recurrence in major depression (see Chapter 47: Deprescribing Benzodiazepines, Stimulants, and Anticonvulsants). After recovery, these mood stabilizers can be tapered off or continued for prevention. Continuation is preferred when there are ongoing stressors; the episodes were severe or recurrent; or the recovery is incomplete.

Second-Line Pharmacotherapy

Antidepressants are second-line strategies for depression with mixed features, and bupropion (Wellbutrin) has the best track record in this class. This antidepressant has a low risk of inducing manic symptoms and may improve depressive mixed features. In two large trials of major depression, mixed features predicted response to bupropion augmentation, although these were secondary analyses (Jha MK et al, *Neuropsychopharmacology* 2018;43(11):2197–2203; Zisook S et al, *Am J Psychiatry* 2019;176(5):348–357).

Trazodone (Desyrel) is another option, with several open-label reports of success from an Italian group (Carmellini P et al, *Int Clin Psychopharmacol* 2025 Jan 13). This antidepressant is often avoided because of its sedative properties, but that can usually be overcome by dosing it entirely at night, even in the higher ranges used for depression (150–300 mg). Outcomes are also improved by titrating trazodone slowly (eg, start 50 mg QHS, raise to 150 mg QHS over one to two weeks). This strategy avoids sudden spikes in trazodone's metabolite, mCPP, which can induce dysphoria.

Lifestyle & Psychotherapy

Psychotherapy plays an important role in depression with mixed features. At the very least, it plays a supportive role, reducing anxiety and preventing treatment dropout. Patients can learn specific skills to manage the symptoms, such as the distress tolerance module from dialectical behavior therapy (DBT). Therapies that focus on sleep (cognitive behavioral therapy for insomnia) and circadian rhythm regulation (social rhythm therapy) are well suited to treating mixed features.

Daily rhythms are harder for these patients to regulate because the manic symptoms tend to predominate in the evening and depressive ones in the morning. Regular sleep times are difficult to implement when they are wired at night and drowsy all day. Instead, start with regular exposure to morning light and evening darkness, such as through a dawn simulator and blue light blocking glasses (both covered in Chapter 14: Therapy and Lifestyle).

As symptoms improve, gradually add in behavioral changes such as consistent sleep and wake times (even on weekends), regular mealtimes, scheduled physical activity (preferably in the morning), and limiting caffeine, especially after noon.

Keep the goals realistic. Mixed features make it difficult for patients to sit still, remember therapy assignments, and control their own thoughts. They may become flooded with anxious thoughts during mindfulness exercises. They may misunderstand complex discussions. If the symptoms are severe, they may do better with shorter sessions, such as 30 minutes a week, instead of the traditional therapeutic hour.

Key Takeaways

- One in four patients with unipolar depression has mixed features.
- They often present in a desperate, dysphoric mood, with anxiety, irritability, agitation, and destructive behavior.
- Antidepressants are risky in mixed features, but you may need to treat the mixed symptoms before trying to taper them.
- SGAs, especially lumateperone and lurasidone, and mood stabilizers are first-line treatments.

CHAPTER 7

Psychotic Depression

Why Psychotic Depression Gets Missed

UNRECOGNIZED BIPOLAR DISORDER MAY BE the most common cause of antidepressant nonresponse, but unrecognized psychosis is just as important. Here are two studies that taught me that.

The first was a reanalysis of STAR*D, a large trial that compared ten treatments across four levels of treatment resistance in depression. Researchers thought that soft signs of bipolar disorder might explain why patients didn't respond to antidepressants in this trial. However, mild manic symptoms and family histories of bipolar disorder were not more common in the nonresponders. Instead, they had soft signs of psychosis, such as mild paranoid thoughts (Perlis RH et al, *Arch Gen Psychiatry* 2011;68(4):351–360).

Perhaps those patients had psychotic features that were missed, and this next study shows us how easy it is to miss them. In 2002, Anthony Rothschild and colleagues organized the largest trial of psychotic depression to date, the STOP-PD study (Study of Pharmacotherapy of Psychotic Depression) sponsored by the National Institute of Mental Health (NIMH). At the end of the trial, they looked back at the hospital records for the patients they enrolled and found the inpatient teams had missed the psychosis in a third of the cases. These were academic centers with a reputation for high standards, but Rothschild's study dealt another blow to that myth. The rate of misdiagnosis was three times higher for the psychiatric attendings than the residents (Rothschild AJ et al, *J Clin Psychiatry* 2008;69(8):1293–1296).

Detecting psychosis can change the treatment plan as much as detecting bipolar. As we'll see, that doesn't just mean adding an antipsychotic.

Hidden Signs of Psychosis

Psychosis comes in two forms: Hallucinations and delusions.

Detecting Hallucinations

Hallucinations are easier to detect than delusions, but asking patients if they ever "hear or see things that other people don't" might miss some cases. Hallucinations involving taste, smell, and bodily sensations are common in psychotic depression. Patients may report feeling touched, foul tastes or smells, electricity in their bones, or that their organs are shifting or their brain is turning into concrete.

Recognizing Delusions

Delusions are more difficult to detect. For the patient, the delusion is not a symptom but the way things are, and they've learned to hide those beliefs from outsiders who don't understand them.

The patient's thought process may tip you off to hidden delusions:

- Speaking in short sentences (poverty of thought)
- Guarded responses
- Rigid, black and white thinking
- Impaired problem solving

They may misinterpret things in an overly literal way (concrete thinking). For example, when you end the session with "See you down the road," and they ask, "What road?"

Look for subtle paranoid themes. Are they guarded when you ask about symptoms, replying with "Why do you ask about that?" Do they ask questions that imply you have malevolent intent? "Did you follow me at the grocery store yesterday?" is a telltale sign, but more subtle breaches of convention can also point the way. "Who were you talking to on the phone before my visit?" asked one patient. Later that night, her family called to tell me she was barricading the doors, afraid that the mafia was out to get her.

Observable Clinical Signs

Perplexed expressions, worried eyes, and a furrowed brow attest to the mental confusion going on beneath the surface. When psychotic depression is severe, patients' skin may appear pale and lifeless, their movements slow and robotic. Their faces are grim, unreactive. They are apathetic and withdrawn, to the point of seeming disconnected from the interview.

Screening for Psychotic Features

The table below provides screening questions that help identify the most common types of depressive delusions. If the patient endorses a delusion, explore how rigid the belief is and how it has affected their behavior. If they believe they have a physical illness, have they gone to doctors or gathered up evidence? Does anyone else agree with their beliefs? Guilt and punishment are common themes in depression, but in psychotic depression, these beliefs become more entrenched and literal. "I deserve death," or even "I am dead," are answers that suggest a psychotic process.

TABLE 7-1. Screening for Psychotic Depression

Delusions	Screening Question
Guilt	Do you feel guilty about anything?
Punishment	Do you feel you are being punished, or that you deserve punishment, like you've done something that can't be forgiven?
Persecution	Do you feel you can't trust people or that they have it in for you? Do you worry that authorities like the police are after you?
Illness	Do you believe you are physically ill? That your symptoms are due to a physical illness?
Nihilism	Does everything seem meaningless? Do you think about death a lot? What are your thoughts about it?
Worthlessness	Do you feel incapable, worthless, or unlovable? That people would be better off without you?
Hallucinations	**Screening Question**
Audio-Visual	Have you heard or seen things that others can't?
Kinesthetic	Have you experienced unusual smells, tastes, or feelings in your body?

Medical Risk and Suicide

While delusions of death and illness are common in psychotic depression, true medical morbidity is also high. These patients have twice the rate of death compared with severe nonpsychotic depression (Vythilingam M et al, *Am J Psychiatry* 2003;160(3):574–576).

Most concerning is the risk of suicide, which is 2 to 5 times higher when psychotic features are present. For these patients, hope has vanished and the pain is unending. They lack the flexibility and problem-solving skills to manage stress. They are out of touch with reality, to the point that they may attempt suicide out of the delusional belief that their family would be better off with the life insurance payout.

Two Types of Psychotic Depression

Conrad Swartz describes two types of psychotic depression: Depression-dominant and psychosis-dominant, depending on which symptoms are more prominent.

1. **Depression-dominant patients** ruminate about depression and are less vocal about the psychosis.
2. **Psychosis-dominant patients** talk obsessively about delusional material. When asked about depressive symptoms, they dismiss them as a normal reaction: "Wouldn't you feel depressed if undercover agents were turning your family against you?" It is tempting to discount symptoms like insomnia and low appetite as normal reactions to an overbearing paranoia, but symptoms have to be assessed in their own right, regardless of what we think about their cause.

Differential Diagnosis

Psychotic depression needs to be distinguished from several other conditions with overlapping presentations:

- **Schizoaffective disorder**—Has psychotic symptoms that persist even when mood symptoms remit.
- **Delusional disorder**—Primary delusions without prominent mood symptoms (difficult to distinguish from psychosis-dominant depression).
- **Dementia with psychosis**—Cognitive deficits are more pervasive and progressive.
- **Bipolar disorder with psychotic features**—Look for history of manic or hypomanic episodes.
- **Psychosis due to a substance use disorder**—Most substances can progress to psychosis, and it is particularly common with recreational use of cannabis, cocaine, ketamine, and stimulants.
- **Psychosis due to a medical disorder**—Most illnesses that affect the brain can cause psychosis, including infections (HIV, syphilis), multiple sclerosis, lupus, tumors, epilepsy, and Parkinson's disease.

Case Vignette: Hidden Psychosis

Julia, a 62-year-old retired teacher, was referred for "treatment-resistant depression" after three antidepressant trials failed. At her first interview, she was slow, expressionless, and spoke in a monotone. When asked directly, she denied hallucinations or delusions. However, she mentioned that she hadn't been eating because "my stomach doesn't work anymore" and that she felt "hollowed out inside." When gently questioned about these beliefs, she revealed that she was convinced her digestive system had "died" and food "just sits there rotting." She didn't tell previous providers because "they would think I'm crazy." After a failed trial of sertraline with olanzapine, she experienced a complete remission with ECT.

Treatment Approaches

Electroconvulsive Therapy (ECT): The gold standard

ECT is the most effective treatment for psychotic depression. Up to 95% of patients with psychotic depression recover fully with ECT (Petrides G et al, *J ECT* 2001;17(4):244–253). That figure, which is based on an observational study of 253 patients, is unheard of in psychiatry. No other intervention has this kind of specificity for psychotic depression; not transcranial magnetic stimulation (TMS), and not ketamine, both of which excluded psychotic depression in their trials.

ECT may be the gold standard, but it's not an easy treatment to introduce. Indecision is the norm in this population, so start the conversation with clarity and optimism: "There is a treatment that can bring you full recovery, so you don't have to suffer like this. It may not be something that you want to do, but it is particularly good for the kind of depression you have. It's not a medication. It is ECT, and nothing else offers you as high a chance of success: as hich as 95%."

Medication Approaches

If you aren't able to arrange ECT, an antipsychotic-antidepressant combination is the next best option, but aim high. The low doses of antipsychotics used for antidepressant augmentation are not likely to work here. In the psychotic depression trials, they were dosed in the ranges used for schizophrenia: olanzapine (Zyprexa) 10–20 mg daily, risperidone (Risperdal) 2–6 mg daily, and quetiapine (Seroquel) 600 mg daily.

Unfortunately, the antipsychotic-antidepressant combo may not move the needle the way ECT does. Antipsychotic augmentation only adds a small benefit compared to antidepressant monotherapy, with an effect size of 0.25 in a meta-analysis of eight trials. Surprisingly, that benefit was for depressive symptoms; the antipsychotic made no detectable difference in psychosis (Farahani A and Correll CU, *J Clin Psychiatry* 2012;73(4):486–496).

Lithium Augmentation

Lithium augmentation can also work in psychotic depression. In open-label trials, 60% of patients with psychotic depression responded to lithium augmentation after an antipsychotic-antidepressant combo failed. Lithium was slightly more effective than haloperidol (Haldol) augmentation in a head-to-head trial, though it was not randomized (Birkenhäger TK et al, *J Clin Psychopharmacol* 2009;29(5):513–515; Ebert D, *J Clin Psychopharmacol* 1997;17(2):129–130). The target level in psychotic depression is not well defined and ranged from 0.4–1.2 mEq/L in these trials.

Choice of Antidepressant

We don't have clear research to tell us which antidepressant to use in psychotic depression. Most trials used tricyclics or selective serotonin reuptake inhibitors (SSRIs), particularly the combination of olanzapine with fluoxetine (Prozac) or sertraline (Zoloft). A few employed mirtazapine (Remeron) or venlafaxine (Effexor) as the antidepressant (Oliva V et al, *Lancet Psychiatry* 2024;11(3):210–220). There is evidence to support the use of bupropion (Wellbutrin), although this antidepressant may induce psychosis in very rare cases (Taipale H et al, *World Psychiatry* 2024;23(2):276–284). The clinical lore is that tricyclics work best in psychotic depression, but this idea hasn't been tested in clinical trials. An older strategy is to use the first-generation antipsychotic loxapine (Loxitane) as monotherapy. Loxapine is metabolized into therapeutic levels of the tricyclic antidepressant amoxapine (Asendin) (Burch EA Jr and Goldschmidt TJ, *South Med J* 1983;76(8):991–995).

Patient and Family Education

Explaining psychotic depression is challenging. Start with areas of agreement, such as distressing symptoms like anxiety or insomnia. Here are some approaches that may help:

- Emphasize that psychotic symptoms are a manifestation of severe depression, not a separate "psychotic disorder."
- Explain that the brain can misinterpret reality under extreme stress.
- Focus on specific symptoms rather than labels: "feeling extremely guilty" or "worrying intensely about health" rather than "delusions."

For patients with rigid delusions, empathize with the stress they are under and avoid directly challenging them.

"It's incredibly stressful to feel unsafe in your own home, and I can see that stress is taking a toll on you. I'm not able to intervene with the police or figure out what is going on, but I can help with the ways that stress affects the brain. If we can improve your energy and concentration, you'll be much better able to manage the stress you're under."

Monitoring Treatment Response

When treating psychotic depression, monitor both mood and psychotic symptoms. Improvement in one domain may not parallel improvement in the other. Some suggestions:

- Use standardized scales for both depression (PHQ-9, HDRS) and psychosis (BPRS)
- Track specific delusional beliefs and their conviction level
- Monitor for emergence of suicidal ideation, especially during the early phase of treatment

Key Takeaways

- Psychotic depression is often missed in practice.
- ECT is the first-line treatment, with remission rates up to 95%.
- Lithium and antipsychotic augmentation are second-line strategies, and antipsychotics usually need to be dosed in the higher ranges used for schizophrenia.

CHAPTER 8

Vascular Depression

VASCULAR HEALTH AFFECTS EVERY ORGAN in the body. When blood flow is occluded, brain cells die within minutes. When large areas are affected, there are noticeable neurologic symptoms, and the sudden injury is called a stroke. When small vessels are blocked, the results are more subtle, but the buildup of these "silent" micro-ischemic injuries can cause depression and cognitive problems.

These vascular injuries cause two types of depression, post-stroke depression and vascular depression. In post-stroke depression, the symptoms start abruptly within the first three months of the stroke. Depression is particularly common after strokes involving the left hemisphere, particularly the prefrontal cortex and basal ganglia (Facucho-Oliveira J et al, *Pract Neurol* 2021;21(5):384–391). In vascular depression, the onset is more gradual, building as the micro-ischemic injuries accumulate. This chapter will focus on vascular depression, but the two syndromes often overlap as both share a common etiology.

Epidemiology and Risk Factors

Vascular depression is common in older adults. After age 50, 1 in 5 patients with depression has a vascular contribution, and this rises to 1 in 2 after age 65 and nearly 100% by age 75 (Taylor WD et al, *Am J Psychiatry* 2018;175(12):1169–1175). Patients with heart disease, diabetes, and hypertension are particularly at risk.

The relationship between vascular disease and depression works both ways. Compromised blood flow leads to neuronal damage in mood-regulating circuits, while mood disorders increase the risk of vascular disease. Cardiovascular disease is more common in mood disorders and starts earlier—six years earlier in major depression and 17 years earlier in bipolar I (Goldstein BI et al, *J Clin Psychiatry* 2015;76(2):163–169).

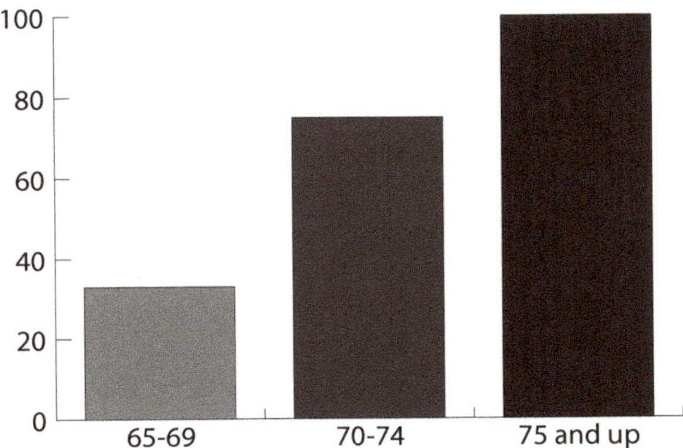

FIGURE 8-1. Age-Related Prevalence of Vascular Depression

Source: Park JH et al, *J Affect Disord* 2015;180:200-206

Clinical Features

The cardinal features of vascular depression are:

- Evidence of small vessel ischemic disease on neuroimaging
- Cardiovascular risk factors
- Cognitive impairment

Outside of those cardinal features, the presentations vary. Often they are indistinguishable from nonvascular depression, but these patients are more likely to have low energy, apathy, and psychomotor slowing. Some have a mix of mood and cognitive symptoms that do not meet the full criteria for a major depressive episode but are nonetheless impairing.

Lack of insight is common, including alexithymia (the inability to describe emotions). An interview with relatives conveys more functional impairment, like trouble making decisions and starting tasks. Executive dysfunction and slow processing speed are common on neuropsychological testing.

TABLE 8-1. Signs of Vascular Depression

Feature	Vascular Depression	Nonvascular Depression
Age of onset	Often later in life	Any age
Cognitive deficits	More prominent	Less prominent
Executive dysfunction	Common	Less common
Psychomotor retardation	More severe	Variable
Insight into illness	Often reduced	Usually preserved
Apathy	Prominent	Variable

TABLE 8-1. Signs of Vascular Depression

Feature	Vascular Depression	Nonvascular Depression
Guilt/worthlessness	Less pronounced	Often prominent
Response to antidepressants	Reduced	Standard

Making the Diagnosis

There are no established criteria for vascular depression, so the diagnosis is made when there is a high suspicion that poor vascular health is contributing to the depressive syndrome. The closest thing to confirmation is "white matter hyperintensities" on an MRI of the brain.* Sometimes these are mentioned in the description even when the summary concludes there are only "age-related changes."

It is unclear whether the number, size, or location of white matter hyperintensities is most critical. The lesions can occur anywhere, but those in the frontal lobes, basal ganglia, and pons have the strongest link to depression (Rushia SN et al, *World J Radiol* 2020;12(5):48–67).

Clinical factors that suggest a vascular contribution include cardiovascular disease, hypertension, diabetes, smoking, and advanced age. The onset of depression later in life without clear psychosocial triggers should raise suspicion, particularly when accompanied by cognitive impairments.

Case Vignette: Vascular Depression

Phil, a 72-year-old man with hypertension and type 2 diabetes, was referred for "treatment-resistant depression" after three selective serotonin reuptake inhibitor (SSRI) trials failed. His symptoms included low mood, apathy, slowed movement and speech, and difficulty concentrating and making decisions. His depression began gradually in his late 60s, with no clear triggers. A brain MRI revealed moderate periventricular and deep white matter hyperintensities. His depression improved with transcranial magnetic stimulation (TMS) and optimization of vascular health.

---

*White matter hyperintensities are best identified with a FLAIR (Fluid-Attenuated Inversion Recovery) MRI. This technique suppresses the signal from cerebrospinal fluid, allowing better visualization of tissue damage.

Treatment Approaches

The main approach is to treat the underlying cause, which means making sure the patient is taking care of their vascular health through medical treatment and lifestyle. Physical activity and a healthy diet (eg, Mediterranean diet, low cholesterol diet, or the DASH diet for hypertension) are a good place to start. Ensuring optimal control of hypertension, diabetes, and hyperlipidemia is essential, as is addressing other vascular risk factors like smoking, alcohol consumption, and obstructive sleep apnea.

After optimizing vascular health, there is good evidence that TMS and electroconvulsive therapy (ECT) are effective in vascular depression (Jorge RE et al, *Arch Gen Psychiatry* 2008;65(3):268–276; Jellinger KA, *J Neural Transm* 2022;129(8):961–976). These treatments are usually more effective than pharmacotherapy.

Pharmacological Options

No antidepressants stand out as particularly effective in vascular depression. Sertraline (Zoloft) has the best safety profile in heart disease, while vortioxetine (Trintellix) may improve cognitive symptoms. Those that can cause hypertension, like venlafaxine (Effexor), are best avoided. Since antidepressant efficacy is greatly reduced in vascular depression, augmentation is often necessary (Aizenstein HJ et al, *BMC Med* 2016;14(1):161).

One promising strategy is nimodipine (Nimotop), an antihypertensive that improves cerebral blood flow. Nimodipine augmentation improved mood in two small controlled trials of vascular depression (Taragano FE et al, *Int J Geriatr Psychiatry* 2001;16(3):254–260; Taragano FE et al, *Int Psychogeriatr* 2005;17(3):487–498). The target dose is 90 mg TID, and it takes up to two months to see an effect. Start at 15 mg TID for patients who are already on blood pressure medicine, or 30 mg TID for other patients, and raise by 15 mg every 10 days. Monitor blood pressure during treatment.

Low-dose methylphenidate (5–20 mg/day) may be helpful, particularly for apathy and cognitive impairment (Mantani A et al, *Am J Geriatr Psychiatry* 2008;16(4):336–337). Medications for dementia are sometimes tried, but this step is best deferred to the neurologists. There are no studies of dementia medications in vascular depression, and their benefits in actual dementia are marginal.

TABLE 8-2. Nimodipine Summary

FDA Approval	Hypertension
Off-Label Benefits	Vascular depression Ultra-rapid cycling bipolar disorder Prevention of cerebral vasospasm after subarachnoid hemorrhage
Dosing	Start at 30 mg TID, raise by 15 mg every 10 days to target of 90 mg TID (If patient is taking an antihypertensive, start at 15 mg TID and consult with PCP)
Risks	Hypotension (check blood pressure and pulse during titration) Avoid in pregnancy (Category C)
Side Effects	Nausea, diarrhea
Half-Life	8–9 hours
Interactions	Take on an empty stomach (food decreases absorption by 30%–50%). Metabolized through CYP3A4 and CYP3A5 with minor contributions from CYP2C19.

Key Takeaways

- Vascular depression is a common cause of treatment resistance in older adults.
- It is characterized by cardiovascular risk factors, cognitive problems, and white matter hyperintensities on a brain MRI.
- Address vascular health, and consider neuromodulation therapies (TMS, ECT) and nimodipine augmentation

CHAPTER 9

Inflammatory Depression

INFLAMMATION IS A NORMAL IMMUNOLOGIC RESPONSE to a perceived attack on the body. That attack can take the form of a wound, infection, the stress response, or any foreign element, from surgical knives to processed foods. When inflammation goes on too long it can cause depression.

Inflammation contributes to one in three cases of depression and one in two cases of treatment-resistant depression (Osimo EF et al, *Psychol Med* 2019;49(12):1958–1970). It is more common in depression with mixed features. Recent surgery, injury, or infection—including COVID-19—are common causes, as are less obvious ones like poor diet, obesity, smoking, chronic stress, chronic medical illness, chemotherapy, postpartum, insomnia, and lack of exercise.

Assessment and Diagnosis

Measuring Inflammation

There are many laboratory markers for inflammation, but the most reliable is the high-sensitivity C-reactive protein (hs-CRP). You need the "high-sensitivity" CRP because a regular CRP is too coarse to detect the lower levels of inflammation that contribute to depression. This low-cost test has more evidence to guide antidepressant therapy than genetic testing.

Signs and Symptoms

Inflammatory depression is not markedly different from general depression, but a few symptoms suggest an inflammatory cause:

- Anhedonia
- Low motivation
- Fatigue
- Brain fog

Look for risk factors for inflammation in the patient's history:

TABLE 9-1. Risk Factors for Inflammation

Childhood trauma	Smoking, sedentary lifestyle
Recent significant stress	Recent chemotherapy or radiation
Chronic medical illness	Recent bodily injury or surgery
Obesity (BMI ≥ 30 kg/m²)	Recent infection
Western diet	Postpartum and perimenopause

References: Ferrucci L and Fabbri E, *Nat Rev Cardiol* 2018;15(9):505–522; Majd M et al, *Front Neuroendocrinol* 2020;56:100800

Treatment

Patients with an hs-CRP ≥ 3 mg/L are less likely to respond to selective serotonin reuptake inhibitors (SSRIs) and more likely to respond to tricyclics (eg, nortriptyline [Pamelor]) or dopaminergics (eg, bupropion [Wellbutrin] and pramipexole [Mirapex]) (Uher R et al, *Am J Psychiatry* 2014;171(12):1278–1286; Jha MK et al, *Psychoneuroendocrinology* 2017;78:105–113; Ventorp F et al, *Psychiatr Res Clin Pract* 2022;4(2):42–47). Inflammation takes a particular toll on the dopamine system, and we see this effect both in neuroimaging and in the clinical interview. These patients often present with fatigue, low motivation, and profound anhedonia, symptoms that point to the dopamine tracks.

Anti-Inflammatory Approaches

Instead of treating the injured dopamine tracks, why not target the inflammation itself? Many medications with anti-inflammatory properties have been tested in depression, including nonsteroidal anti-inflammatory drugs (NSAIDs), glucocorticoids, cytokine inhibitors, statins, minocycline (Minocin), and pioglitazone (Actos). Among them, the NSAID celecoxib (Celebrex) has the best evidence of efficacy.

Unfortunately, few trials of anti-inflammatory agents have tested whether these treatments work better in patients with high levels of inflammation. Celecoxib is one exception, and another is minocycline. As we'll see in Chapter 26: Minocycline, this antibiotic works preferentially in patients with an elevated hs-CRP.

Some complementary therapies have anti-inflammatory properties, and three of them have greater efficacy in depression when inflammation is high: n-acetylcysteine (NAC), omega-3 fatty acids, and possibly probiotics

(Porcu M et al, *Psychiatry Res* 2018;263:268–274; Milajerdi A et al, *Eur J Nutr* 2020;59(2):633–649). L-methylfolate (Deplin) is also more effective here, possibly because inflammation impairs folate metabolism (Shelton RC et al, *J Clin Psychiatry* 2015;76(12):1635–1641). In the case of omega-3, inflammation also guides the dose. Patients with inflammatory depression were twice as likely to respond to high doses of omega-3 (EPA 4,000 mg/day) than the standard dose range (EPA 1,000–2,000 mg/day) in a small trial (Mischoulon D et al, *J Clin Psychiatry* 2022;83(5):21m14074).

The Obesity Connection

Obesity and depression are linked, and not just because of the stigma surrounding weight. The foods that cause obesity — fast food, fried food, processed and sugary foods — also cause inflammation, as do the fat cells themselves. Fat cells grow quickly and outstrip their blood supply, which causes those cells to die. The necrotic cells trigger an inflammatory response.

Some treatments that work better in inflammatory depression are also more effective in depression with obesity (BMI ≥ 30 kg/m^2), such as augmentation with bupropion, l-methylfolate, and omega-3s.

Lifestyle and Health

Besides depression, inflammation raises the risk of heart disease, diabetes, and cancer. A visit to the PCP is in order when the hs-CRP is high, and these patients need an extra dose of healthy lifestyle. Exercise, Mediterranean diet, sleep hygiene, and stress reduction techniques like mindfulness all reduce inflammation.

Treatment: A Step-By-Step Approach

1. Measure hs-CRP in patients with treatment-resistant depression
2. If hs-CRP ≥ 3 mg/L, consider changing the medication strategy
3. Address lifestyle factors that contribute to inflammation
4. Aim for symptomatic recovery, but monitor hs-CRP every 6–12 months to see if it improves (optimal level is below 1 mg/L)

The table lists treatments to consider for inflammatory depression, and we'll have more details on how to use them in future chapters.

TABLE 9-2. Treatments for Inflammatory Depression

First-Line Medications	
Celecoxib	This anti-inflammatory pain medication has 17 randomized trials in depression, and a few of them suggest it works preferentially when inflammation is high (200 mg BID).
Bupropion	Inflammation (CRP ≥ 3 mg/L) and obesity (BMI ≥ 30 kg/m2) predict response to bupropion augmentation of SSRIs (150–450 mg XL QAM).
Pramipexole	This medication targets the dopamine tracks that are disrupted by inflammation, and it reduced inflammatory markers and anhedonia in trials of depression (0.5–2.5 mg QHS).
Second-Line Medications	
Nortriptyline	Elevated CRP predicts a better response to nortriptyline than escitalopram in major depression (50–150 mg/day, dose by serum level).
Minocycline	This antibiotic has anti-inflammatory properties. It has a mix of positive and negative results in depression, but trends toward the positive in patients with elevated CRP (100 mg BID).
Complementary and Alternative Therapies	
N-acetylcysteine (NAC)	Elevated CRP predicts response to NAC in anxiety and depression (2,000 mg/day).
L-methylfolate	Inflammation impairs folate metabolism, and elevated inflammation (CRP ≥ 3 mg/L) predicts response to this folate vitamin (15 mg/day).
Omega-3	Elevated CRP predicts response to omega-3, and higher doses may be needed for inflammatory depression (4,000 mg EPA).
Lifestyle	Exercise, yoga, tai chi, the Mediterranean diet, mindfulness, and cognitive behavioral therapy for insomnia have anti-inflammatory and antidepressant effects.

Patient Education

Patients often find the inflammatory model validating. Frame it simply: "Your blood test shows that inflammation is affecting your brain and your body. This explains why you're experiencing these symptoms and why standard antidepressants haven't worked well. The good news is we can target this inflammation directly."

Emphasize that addressing inflammation improves both mental and physical health and connect that to the benefits of lifestyle change. "You know, as a doctor, I advise most patients to exercise, but from what I see in your labs, exercise is going to be particularly important to your mood and health." To enhance motivation, track their progress by measuring hs-CRP around every six months.

Case Vignette: Inflammation after Surgery

Huan is a 48-year-old man with obesity who developed depression after knee surgery. Although the operation was successful, his sertraline "stopped working completely" afterward. Lab tests showed an hs-CRP of 4.2 mg/L. Switching to bupropion and adding celecoxib improved his symptoms within three months, and incorporating Mediterranean diet and exercise further reduced his hs-CRP to 1.8 mg/L over six months with ongoing mood improvement.

Key Takeaways

- Inflammation wears down physical and mental health, contributing to 50% of cases of treatment-resistant depression. It is measured with an hs-CRP (≥ 3 mg/L).
- Preferred treatments for inflammatory depression include dopaminergic medications (bupropion, pramipexole), tricyclics, anti-inflammatories (celecoxib, minocycline), and complementary therapies (NAC, l-methylfolate, omega-3s).
- Lifestyle interventions that target inflammation are a core part of treatment.

CHAPTER 10

Trauma and Depression

TRAUMA IS COMMON IN DIFFICULT-TO-TREAT depression, so common that it is unfair to call it a unique subtype. Childhood trauma is a common pathway for 75% of patients with chronic depression, and 50%–70% of those with treatment-resistant depression have survived traumatic experiences (Negele A et al, *Depress Res Treat* 2015:650804). For these patients, depression is more severe, more chronic, and starts earlier.

Undisclosed Trauma

Trauma hides behind a rock. Even direct questions may not get to a trauma history that is covered in shame and avoidance. When asking about trauma, let patients know that they don't need to go into the details if they are not comfortable, but that knowing whether they experienced trauma can help guide their treatment. Some patients answer more honestly on paper, and I've created a screening instrument to gather that history (modified from several adverse experience scales):

Traumatic Life Events

Below are difficult or stressful things that sometimes happen to people. As you look through them, think about your life, including your childhood, and check as many boxes as apply.

SCREENER 10-1. Trauma and Depression Screener

Event	Happened to Me	Witnessed It	Learned About It	Not Sure	Doesn't Apply
Natural disaster (*examples:* flood, hurricane, earthquake, fire, explosions)					
Accidents and toxicity (*examples:* car accident, explosions, exposure to dangerous chemicals)					

SCREENER 10-1. Trauma and Depression Screener

Event	Happened to Me	Witnessed It	Learned About It	Not Sure	Doesn't Apply
Assault (*examples:* attacked, hit, slapped, kicked, beaten up, shot, stabbed, threatened)					
Sexual assault (rape, attempted rape, made to perform any type of sexual act through force or threat of harm, or any other unwanted or uncomfortable sexual experience)					
Combat or exposure to a war-zone (in the military or as a civilian)					
Captivity (*examples:* kidnapped, abducted, held hostage, prisoner of war)					
Life-threatening illness, injury, or severe suffering					
Sudden, unexpected death of someone close to you					
Any other very stressful event or experience					

As a child . . .	Yes	Not Sure	Doesn't Apply
Did you feel that you didn't have enough to eat, had to wear dirty clothes, or had no one to protect or take care of you?			
Did you live with anyone who was depressed, mentally ill, had a problem with drugs, or attempted suicide?			
Did your parents or adults in your home ever hit, punch, beat, or threaten to harm each other?			
Did you live with anyone who went to jail or prison?			
Did a parent or adult in your home ever swear at you, insult you, or put you down?			
Did a parent or adult in your home ever hit, beat, kick, or physically hurt you in any way?			
Did you feel that no one in your family loved you or thought you were special?			
Were you frequently bullied by other children at school?			

When patients describe intense trauma, we may feel overwhelmed, unsure of what to say. A simple empathetic response goes a long way: "I'm sorry that happened to you."

Punishment, Avoidance, and Depression

Childhood trauma leads to patterns of avoidance that contribute to chronic depression. As one patient explained, "At a young age, I came to understand that I would never get the love and support I needed from my parents. Instead, it was anger, derision, or just acting like I didn't exist (which I think they would have preferred). So I withdrew into myself. I made up stories and spent many days alone in my room. When I broke my arm, I didn't tell anyone. It was safer that way."

These patients learned to navigate the world by escaping punishment, rather than seeking out rewards. The result is avoidance and anxiety, particularly toward people who remind them of their parents: authority figures, intimate relationships, and caregivers. While avoidance helped them survive a chaotic childhood, it has outrun its purpose in their adult life, where it contributes to depression and isolation.

Behavioral patterns that are shaped by punishment are much more generalized and ingrained than those shaped by reward. That means that change is difficult and slow for these patients, particularly when the change involves interacting with other people.

Trauma and Chronic Depression

Psychologist James McCullough developed a psychotherapy for chronic depression that addresses the effects of early childhood trauma. The therapy, which goes by the unwieldy name of cognitive behavioral analysis system of psychotherapy (CBASP), is a pragmatic blend of behavioral and psychodynamic approaches with good evidence in this population.

In McCullough's view, patients with chronic depression have trouble changing because they are unable to take in feedback from the outside world. The patient is "isolated interpersonally, talks in a monologue manner using a well-rehearsed script of rejection, and lives in quiet despair within a self-contained world that is not informed by external influences. Nothing new enters and nothing leaves this phenomenological orbit. The patient presents with a terrible sense of 'sameness.' Existentially, the patient describes a lifestyle where time appears to have stopped—the present reflects the past and the future bodes only more of the same." (McCullough JP Jr, *Front Psychiatry* 2021;11:609954).

It is as if they are encased by a concrete wall. McCullough theorizes that they put up this concrete during childhood to protect themselves from the pain of emotional mistreatment, physical abuse, sexual abuse, neglect, and parental loss, problems that three out of four patients with chronic depression have lived through. Part of the work is to help them regain a sense of agency in their own lives, to slowly bring down the concrete wall. At first, the newfound awareness is painful, as the patient realizes how their own actions are causing their suffering. "If you don't like the way you feel, then you must change your behavior," is the message McCullough puts front and center.

Post-Traumatic Stress Disorder (PTSD) vs Depression with Trauma

Trauma can cause depression, PTSD, or both disorders. For patients with PTSD, the world is dangerous, uncontrollable, and unpredictable. They are hypervigilant, scanning their environment and going out of their way to avoid triggering reminders. All this effort makes them tense and exhausted. They have nightmares, flashbacks, and are easily startled.

Besides PTSD, trauma causes two other disorders that often overlap with depression:

1. Complex PTSD: Chronic PTSD and relationship problems that stem from extensive interpersonal trauma
2. Prolonged grief disorder: Symptoms of grief that persist longer than a year, often after a traumatic loss

Complex PTSD

These patients have lived through interpersonal trauma that is prolonged and extensive, like abuse, captivity, prolonged domestic violence, or torture. To make this diagnosis, the patient first has to meet criteria for PTSD. Then, on top of it, they have to meet a few criteria that describe difficulties in relationships, like difficulty feeling close to others; problems in identity, like shame and low self-worth; and problems in emotion regulation, like overreacting to everyday stress, violent outbursts, or self-destructive behaviors. The diagnosis was added to ICD-11 in 2018 but has not made it into the DSM.

Psychotherapy is the main treatment for Complex PTSD, particularly trauma-focused therapies and dialectical behavioral therapy (DBT). There are no medication trials in this syndrome.

Prolonged Grief Disorder

Prolonged grief disorder entered DSM in 2022 and describes significant grief that lasts longer than a year. Often the grief is sudden and traumatic, and the symptoms resemble those of PTSD. These patients avoid reminders of the deceased, cannot stop thinking about them, and feel numb and detached, "like an automaton, putting on a mask."

Although depressive symptoms are a common part of this syndrome, it does not respond well to antidepressants. In a large, randomized controlled trial, psychotherapy treated prolonged grief disorder while a selective serotonin reuptake inhibitor (SSRI) did not, even though two-thirds of the patients also had depression (Shear MK et al, *JAMA Psychiatry* 2016, 73(7):685–694). The therapeutic approach integrates exposure-based components from trauma-focused therapy. For example, in complicated grief treatment the patient records the narrative details of the death and listens to the recording once a day.

Antidepressants and Trauma

Many clinicians have the impression that patients with trauma histories respond better to psychotherapy than antidepressants, but this is difficult to prove. Antidepressants and psychotherapy work equally well when there is a history of trauma, regardless of the type of trauma, and regardless of whether they have depression alone or depression with PTSD (Childhood Trauma Meta-Analysis Study Group, *Lancet Psychiatry* 2022;9(11):860–873; Green BL et al, *J Clin Psychol* 2006;62(7):815–835). However, patients tend to respond best to the treatment modality they prefer, and that is a more useful guide.

Knowing that the depression developed from a traumatic background does not tell us which antidepressant is likely to work unless they also have significant symptoms of PTSD. In that case, it is best to start with an antidepressant that also works in PTSD. Those with the best evidence are SSRIs and serotonin-norepinephrine reuptake inhibitors (SNRIs), specifically:

- Sertraline (Zoloft) (50–200 mg/day)

- Paroxetine (Paxil) (20–50 mg/day)
- Fluoxetine (Prozac) (20–40 mg/day)
- Venlafaxine (Effexor) (75–225 mg/day)

However, these antidepressants have limitations. They are more effective for symptoms of depression than PTSD, and they generally failed in trials of combat-related PTSD. Another strategy is to combine sertraline (100–200 mg/day) with the antipsychotic brexpiprazole (Rexulti) (1–3 mg/day). In several industry-sponsored trials, this combination was more effective than either alone (Hobart M et al, *J Clin Psychiatry* 2025;86(1):24m15577).

Hyperarousal and Mixed Features

When a depressed patient feels anxious, irritable, impulsive, keyed up, and on edge, are they suffering from mixed features or the hyperarousal of PTSD? The two syndromes look similar, and both can be true at the same time. Trauma makes mixed features more likely, both in unipolar and bipolar depression (Frazier EA et al, *Child Psychiatry Hum Dev* 2017;48(3):393–399). To tell them apart, look for nonsymptomatic signs of bipolar, like family history and treatment response.

Mixed features and hyperarousal both get in the way of sleep, and they improve when sleep is treated.

Insomnia

Whether from hyperarousal, mixed features, or ruminative thoughts, insomnia is common in patients with a trauma history. Here are common ways that insomnia presents in these patients:

- **Trouble falling or staying asleep**—Does the bedroom or nighttime trigger traumatic memories? Do they dread the repetitive nightmares that come with sleep? Do they sleep with the TV on to drown out these fears (if so, recommend sleep music or audio books to avoid the sleep-disrupting effects of blue lights described on pages 104–105).
- **Trouble staying asleep**—Do they wake up in a state of alarm and pace around the room with worry and fear, unable to return to sleep?
- **Nightmares**—These can occur with or without PTSD. Ask how the nightmares affect their mood the next day and how frequent they are. If the answer is at least once a week, the nightmares are probably affecting their mental health.

- **Night terrors**—This is when patients scream, kick, or thrash about during sleep, or sleepwalk. Patients may not remember these episodes, but their spouse usually will.
- **Sleep apnea**—The rate of sleep apnea is high in PTSD (40%–80%). These patients are often young and thin, and don't resemble the Pickwickian stereotype of a middle-aged, obese man. Sleep apnea, in turn, raises the risk of depression, and unless it is treated the patient will not respond as well to antidepressants.

Mood improves when the underlying sleep problem is treated. That treatment may involve CBT-insomnia, a CPAP or BiPAP device for sleep apnea, a hypnotic like eszopiclone (Lunesta), or prazosin (Minipress). David Osser consults on treatment-resistant cases and finds particular success with prazosin when trauma and depression overlap.

Prazosin

Prazosin is an alpha-1 blocker antihypertensive that improves nightmares and sleep quality in PTSD. Prazosin also improves daytime PTSD symptoms, like anxiety, anhedonia, and hyperarousal (Reist C et al, *CNS Spectr* 2021;26(4):338–344). Prazosin may even treat depressive symptoms, either directly or by improving sleep quality. In a randomized trial, low-dose prazosin (0.5–1 mg QHS) augmented antidepressants in patients with depression and a history of trauma, although most of the patients also suffered frequent nightmares (Guo P et al, *Acta Psychiatr Scand* 2025; 151(2):142–151).

TABLE 10-1. Prazosin Summary

FDA Approval	Hypertension
Off-Label Benefits	PTSD (nightmares and daytime symptoms) Benign prostatic hyperplasia
Dosing	Start 1 mg QHS, raise by 1–2 mg every 4–7 days based on response. Average 12–16 mg/day; max 25 mg/day for men and 12 mg/day for women. Divide the dose at ≥ 3–5 mg/day (give 25% in morning).
Risks	Syncope (1%, usually occurs in first week) Check blood pressure and pulse and monitor for falls during titration. Consult with PCP if patient is taking other antihypertensives. Avoid in pregnancy (Category C)

TABLE 10-1. Prazosin Summary

Side Effects	Dizziness, headache, fatigue, palpitations, nausea
Half-Life	2–4 hours
Interactions	Additive hypotension with other antihypertensives

Key Takeaways

- Depression that develops out of a traumatic background tends to be more severe and chronic.
- While no antidepressant stands out as more effective in traumatic depression, addressing sleep problems like insomnia, nightmares, and sleep apnea can speed recovery.
- Depression is a common symptom of several trauma-related disorders: PTSD, Complex PTSD, and—in the case of traumatic loss—prolonged grief disorder. These disorders respond better to psychotherapy than medication.

CHAPTER 11

Other Subtypes of Depression

PSYCHOSIS, INFLAMMATION, AND VASCULAR DISEASE are common causes of antidepressant resistance. The subtypes in this chapter respond to antidepressants but may require some fine-tuning of the strategy.

Atypical and Melancholic Features

Atypical and melancholic depression have opposite symptom profiles. While the atypical oversleeps and overeats, the melancholic wakes up early and has little appetite. In atypical depression, emotional energy is focused outward. They cheer up when good things happen and become tearful and dejected at the slightest rejection. In contrast, the melancholic is pulled inward, as if by a magnet. They stew with guilty ruminations about the past and groundless fears about the future.

Melancholic features are evenly distributed between the sexes, while atypical features are more common in women, particularly women of childbearing age. Patients with melancholic and atypical depression differ in their baseline personalities, though these differences are not absolute.

Melancholic depression often arises as if out of nowhere in patients with a healthy personality and no prior psychiatric history (Valerio MP et al, *J Nerv Ment Dis* 2020;208(10):810–817). Atypical depression, in contrast, is more common in people with anxious and neurotic traits. People with neuroticism tend to experience negative emotions under stress, like anger, anxiety, self-consciousness, irritability, and depression. They interpret ordinary situations as threatening and minor frustrations as hopelessly overwhelming.

Despite its name, atypical depression is common, particularly in outpatient settings. The misnomer dates back to its discovery in the 1950s, when depression was rarely recognized outside of the hospital and melancholic

TABLE 11-1. The Neurotic Temperament

I get stressed-out easily
I dislike myself
I worry about things
I have frequent mood swings
I often get angry or irritated
I panic easily
I am often down in the dumps
I am easily bothered by things
I have trouble relaxing
I am easily frustrated
I fear the worst
I am rarely calm under pressure
I feel threatened easily
I am filled with doubts about things

types predominated in the wards. That origin story also points to a pharmacologic difference between them.

The antidepressant era was launched with the tricyclic imipramine (Tofranil) in 1957 and a monoamine oxidase inhibitor (MAOI; iproniazid) in 1958. At first, they were tested in hospitalized patients. In those trials, the tricyclics looked more effective, but the inpatient populations were enriched with melancholic depression (also called *endogenous* or *involutional* depression).

Two psychiatrists arrived at different results. Working at an outpatient mood clinic in London, Drs. ED West and PJ Dally saw a distinct pattern as they tested the MAOI in 500 outpatients. MAOI responders tended to be more anxious and neurotic, and had somatic symptoms that were the reverse of those typically seen in hospitalized patients. They coined the term atypical depression to describe them (West ED and Dally PJ, *Br Med J* 1959;1:1491–1494).

The original MAOI, iproniazid, was taken off the market a few years later due to liver toxicity, but the close connection between atypical features and MAOIs continued, buoyed by studies from Donald Klein's group at Columbia University. In those trials, patients with atypical depression were more likely to respond to phenelzine (Nardil, 70%) than imipramine (40%) (Pae CU et al, *CNS Drugs* 2009;23(12):1023–1037).

In contrast, melancholic depression is more responsive to tricyclic antidepressants, although some studies question this advantage (Perry PJ et al, *J Affect Disord* 1996;39(1):1–6; Undurraga J et al, *J Psychopharmacol* 2020;34(12):1335–1341). The melancholic type may also be more responsive to electroconvulsive therapy (ECT), but this association has also come under question. Instead, older age, greater severity, and psychotic features predict ECT response. Those features overlap with melancholia, which tends to have a later onset and greater severity than atypical depression (Veltman EM et al, *J ECT* 2019;35(4):231–237).

TABLE 11-2. Opposing Symptoms: Melancholic vs Atypical Depression

Feature	Atypical	Melancholic
Mood quality	Tearful, anxious, relatable in a way that is similar to ordinary sadness	Different from ordinary sadness: anhedonic, empty, despondent, tormented by groundless fears
Timing and reactivity	Worse in evening; highly reactive to daily events, especially rejection	Worse in morning; does not react much to daily events
Thoughts	Easily overwhelmed by stress and interpersonal rejection	Inflexible negative thoughts of guilt, nihilism, worthlessness, or foreboding doom, which may reach psychotic level
Appetite	High	Low
Sleep	Oversleeping	Early morning awakening
Motor	Leaden paralysis (arms or legs feel pulled down by lead weights)	Marked psychomotor agitation (eg, pacing) or retardation (eg, slow, unmoving)
Population	More common in outpatients, women (before menopause), and bipolar II	More common in inpatients

Anxious Depression

Anxiety is a synonym for distress, and as such is a poorly defined symptom. It is common to nearly all psychiatric diagnoses but diagnostic of none. Three-quarters of patients with depression have significant anxiety (Hasin DS et al, *JAMA Psychiatry* 2018;75(4):336–346), and common causes of this anxious depression are listed in the table.

Some of those causes point toward specific treatments. Neurotic temperament predicts a favorable response to selective serotonin reuptake inhibitors (SSRIs), atypical depression to MAOIs, and melancholic to tricyclics.

TABLE 11-3. Causes of Anxious Depression

Cause	Features	Pharmacotherapy
Anxiety disorder	The anxiety is accompanied by phobic behavior or, in the case of generalized anxiety, futile attempts to stop the worry.	SSRI or SNRI (dosed at the higher end of the optimal ranges for depression, see table 15.1).
Atypical features	Anxiety is a reaction to rejection or daily stressors	MAOIs or SSRIs
Melancholic features	Anxiety about guilt or a foreboding sense of doom	Tricyclics
Mixed features	Free floating nonspecific anxiety with somatic symptoms	Mood stabilizers and second-generation antipsychotics (SGAs)
Neurotic temperament	Lifelong tendency to experience negative emotions, such as anxiety, self-consciousness, and irritability	SSRIs
Rumination	A repetitive cycle of negative self-talk	Does not point to a specific med, but is a risk factor for recurrence of depression
Distress	Some patients use "anxiety" as a synonym for distress	Does not point to a specific med, but is a risk factor for side effects, treatment dropout, and suicide

Psychotherapy is an important part of the plan, in part because comorbid anxiety disorders respond better to psychotherapy, but also because these patients have a higher risk of dropping out of treatment and benefit from more support. This is particularly true for patients with DSM-5's anxious distress specifier.

This specifier recognizes patients with high levels of nonspecific anxiety and at least two of these symptoms:

- Keyed up or tense
- Restless
- Trouble concentrating because of worry
- Sense of impending doom, as if something awful may happen
- Fear that they might lose control of themselves

The anxious distress specifier tells us little about the treatment but much about the management. These patients are more likely to have side effects and less likely to recover fully. They have low frustration tolerance and are more likely to drop out of treatment or attempt suicide. They benefit from closer monitoring, outreach between sessions, and faster acting therapies such as the ketamines, dextromethorphan-bupropion combination (Auvelity), or antipsychotic augmentation.

What about benzodiazepines for anxious depression? As we'll see in Chapter 28: Alprazolam and the Benzodiazepines, they are robustly anxiolytic and there is good evidence that they speed up antidepressant response, particularly alprazolam (Xanax). A popular strategy from the 1990s was to start a benzodiazepine with an antidepressant with the plan to taper the benzodiazepine off as the antidepressant took effect. This works well but has faded in popularity due to the difficulty of stopping the benzodiazepine. With only one in three patients reaching full recovery on an antidepressant, the odds are stacked toward long-term benzodiazepine use. Although benzodiazepines help depression in the short term, observational studies suggest a risk of worse cognition, problem-solving abilities, and depression after years of use.

Benzodiazepines are anxiolytics, and they deserve that classification. I have never seen a patient develop paradoxical anxiety after starting them. That can happen with antidepressants, including the SSRIs, which are not directly anxiolytic. SSRIs are preferred for anxiety disorders and neurotic temperament but otherwise do not have an edge in anxious depression. Bupropion (Wellbutrin) is just as effective as SSRIs in anxious depression, and no more likely to cause anxiety, based on multiple, large, head-to-head trials. Although bupropion works in anxious depression, whether it treats anxiety disorders is less clear. High quality studies there are lacking (Bystritsky A et al, *Psychopharmacol Bull* 2008;41(1):46–51).

Mirtazapine (Remeron) augmentation may have a role in anxious depression. This strategy gained popularity in the 2000s, buoyed by the clever nickname "California rocket fuel." The early, small trials were promising, but the strategy ran out of fuel when it was tested in three large randomized trials (Jordan T and Aiken C, *Carlat Psychiatry Rep*, August 30, 2019). However, it did work in a subset of patients: those with high anxiety (Rifkin-Zybutz R et al, *J Psychopharmacol* 2020;34(12):1342–1349).

Buspirone (BuSpar) is another popular augmentation strategy that is sometimes called on for anxious depression. Like mirtazapine, buspirone's augmentation trials are largely negative, but it may have a role in depression with high levels of anxiety and low levels of treatment resistance. In trials from the 1990s, buspirone worked as monotherapy in this population, reducing both depressive and anxious symptoms (45–90 mg/day, divided BID or TID) (Fulton B and Brogden RN, *CNS Drugs* 1997;7:68–88).

A more reliable medication for anxious depression is quetiapine (Seroquel), but its low tolerability means it is best reserved for severe cases. Quetiapine augmentation relieved anxiety and depression in two randomized controlled trials of major depression with high anxiety (average dose 200 mg QHS). As monotherapy, quetiapine lowers anxiety in bipolar disorder (300 mg QHS) and treats generalized anxiety disorder (GAD) (50–150 mg QHS).

Seasonal Depression

Lack of sunlight in the winter months can cause depression. Some people are more sensitive to the rapid decline in sunlight around the equinox in late September. Others are triggered by the "bleak midwinter." Their depression begins as the light reaches a nadir in late December and continues into January and February.

If this pattern happens two years in a row, it is recognized in DSM-5 as a seasonal pattern. This specifier can also apply to bipolar disorder, as long as there is a consistent match between specific seasons and particular moods, most often manias or mixed states in spring and depressions in winter.

Seasonal moods are closely linked to latitude. In the US, the prevalence of winter depression falls from 10% in New Hampshire and Alaska to 6% in Maryland. It virtually disappears in the tropics (around 1% in Florida), and then starts to rise as we get further from the equator (in Australia, the highest rates are in the Southernmost region, Tasmania) (Rosenthal NE, *Winter Blues*, Guilford Press, 2012; Nevarez-Flores AG et al, *J Psychiatr Res* 2023;162:170–179).

Atypical features often go along with winter depression. These patients sleep more, eat more, and crave carbohydrates, suggesting an evolutionary link to animal hibernation. Seasonal depression can also have a summer pattern, and here we see a reversal of those somatic symptoms. Patients with summer depression sleep less, eat less, and generally resemble the melancholic subtype. Summer depression is more common in China, Japan, and the Philippines, for reasons that are not well understood.

In practice, patients with seasonality are often unaware of their own seasonal pattern. Instead, they tend to attribute it to life events, such as holiday stress, returning to school or, in adulthood, getting their kids back to school. Some notice the pattern after an abrupt change, such as moving north or into a house with fewer windows.

Seasonal winter depression is one disorder where the treatment directly addresses the cause: light therapy (Chapter 36: Light Therapy). The syndrome also responds to aerobic exercise, cognitive behavioral therapy (CBT), and—if it is not a bipolar type—antidepressant therapy. Both fluoxetine (Prozac) and bupropion have good support, but only bupropion has FDA approval in seasonal affective disorder, where it is approved for prevention of fall and winter episodes. Vitamin D deficiency may also play a role, as its levels decline with lack of sunlight, but the studies here are less consistent.

All the treatments I just mentioned (including light therapy) work in general depression as well, so what does recognizing the seasonal pattern add to the picture? Prevention. If we know when the patient's depression typically begins, we can plan ahead, for example, by starting light therapy two weeks before the typical onset. Some patients take a low dose of bupropion (eg, 150 mg) throughout the spring and summer, and then raise it to 300 mg as the winter nears.

Postpartum Depression

Postpartum depression begins within four to six weeks of delivery, with a peak onset at two weeks. Until recently, postpartum depression was treated like any other episode of major depression, with psychotherapy and antidepressants; sertraline (Zoloft) is the favored agent because it is relatively safe during breastfeeding. Two developments have changed that approach.

First is the recognition that many cases of postpartum depression are due to bipolar disorder. This is true even when it is the first depressive episode in the woman's life, and especially true if the episode is severe or involves mixed or psychotic features (Munk-Olsen T et al, *Arch Gen Psychiatry* 2009;66(2):189–195). This makes antidepressants riskier in the postpartum period.

The other sea change is the approval of the neurohormone zuranolone (Zurzuvae) for postpartum depression. Zuranolone is a synthetic version of allopregnanolone, a hormone whose precipitous fall after delivery contributes to postpartum depression. Compared to SSRIs, it is definitely faster and possibly more effective, but it comes with limitations that we'll discuss in Chapter 33: Zuranolone.

Hormonal shifts are not the only cause of postpartum depression. It is an inflammatory state, and the anti-inflammatory celecoxib (Celebrex) is effective in postpartum depression. Circadian rhythms are disrupted, and evidence supports the use of morning light therapy and evening darkness (or blue light blocking glasses when awakened by the newborn) (Bennett S et al, *Med Hypotheses* 2009;73(2):251–253). For women who undergo a cesarean section, using esketamine (Spravato) as the anesthetic may prevent postpartum depression (Frivaldszky L et al, *J Psychiatr Res* 2025;183:164–173).

Exercise and a healthy diet are helpful for postpartum episodes, and psychotherapy is a critical part of recovery. As a first step, bring the family in to improve understanding, reduce conflict, and problem-solve ways to increase support and promote the mother-infant bond. Simply providing the mother with an ergonomic infant carrier prevented postpartum depression in a randomized trial of low-income mothers (Little EE et al, *J Affect Disord* 2023;340:871–876).

Other Subtypes

Other subtypes of depression have been proposed based on the predominant symptoms, like fatigue or cognitive impairment. Potential approaches for these are listed in the table.

TABLE 11-4. Augmentation Strategies for Depressive Subtypes

Melancholic features	Tricyclics
Atypical features	MAOIs
Anxious depression	Mirtazapine (30–45 mg QHS) Buspirone (45–90 mg/day, divided BID-TID) Quetiapine (150–300 mg QHS)
Mixed features	Aripiprazole (5–10 mg QD) Lumateperone (42 mg QHS) Lurasidone (20–60 mg QD with food) Ziprasidone (40–80 mg BID with food) Bupropion (150–300 mg XL Qam) Trazodone (150–300 mg QHS)
Postpartum depression	Zuranolone (50 mg QHS)

TABLE 11-4. Augmentation Strategies for Depressive Subtypes

Seasonal depression	Light therapy
Fatigue	Modafinil (100–200 mg Qam) Armodafinil (150–250 mg Qam)
Insomnia	Eszopiclone (3 mg QHS)
Cognitive impairment	Vortioxetine (10–20 mg QD) Bupropion (150–300 mg XL Qam) TMS Probiotics
Suicidality	Lithium (target level: 0.6–0.8 mmol/L)

Key Takeaways

- Depressive subtypes can guide treatment selection, such as MAOIs for atypical features, tricyclics for melancholic features, light therapy for seasonal depression, and zuranolone for postpartum depression.
- Anxious depression has many causes, and each points the way to a different intervention.
- Treatment for postpartum depression has evolved with greater recognition of the risk for bipolar disorder in this period and therapies like zuranolone.

SECTION III

Treatment

Psychosocial Treatment

CHAPTER 12

The Psychology of Depression

HOPELESS, UNMOTIVATED, FORGETFUL. The symptoms of depression get in the way of recovery. Clinicians need to be on alert for these problems, gently pointing them out when they get in the way of care. Ideally, patients will see that it is their depression acting up when, for example, thoughts of worthlessness cause them to miss appointments. In chronic depression, the symptoms can weave their way into personality. For these patients, depression is who they are, not what they have. Gaining perspective on their symptoms is difficult, but not impossible.

Hopelessness

Patients often give up on lifestyle change or neglect to fill their medication prescriptions out of hopelessness, an overarching feeling that nothing will work. Optimism is the antidote, but unless it is balanced with realism, it can backfire. Excessive optimism inspires mistrust, especially among those with chronic depression.

Hopelessness is contagious. I keep a long list of options for depression on my desk to ensure that I don't fall into the countertransference trap of giving up and doing nothing.

Passivity

Passivity begins from the moment the patient enters the room and asks where you'd like them to sit. It extends to medication choices and includes patients who look to medications for recovery at the expense of therapy and lifestyle change. Although biological explanations of depression help patients accept medication, they can also foster passivity. In experimental studies, patients become more pessimistic about treatment when told that their depression is biological rather than psychological in origin. Biological explanations are also stigmatizing, causing others to distance themselves

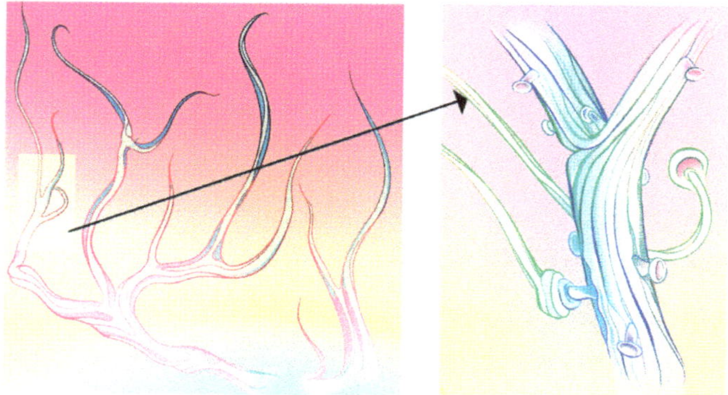

FIGURE 12-1. Stress and depression have caused the neurons in this picture to shrink back, forming fewer dendritic connections.

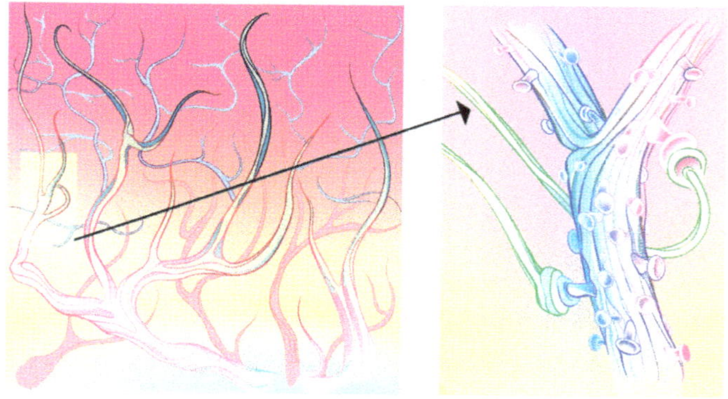

FIGURE 12-2. The neurons have grown and formed more connections after treatment with neuroprotective agents like medication, psychotherapy, exercise, and other lifestyle changes.

and view patients as dangerous (Kemp JJ et al, *Behav Res Ther* 2014;56:47–52; Loughman A and Haslam N, *Cogn Res Princ Implic* 2018;3(1):43).

One way to overcome this dilemma is to emphasize how lifestyle changes the brain in ways that synergize with antidepressants. Exercise, Mediterranean diet, sleep regulation, socializing, and stress reduction all reshape the brain, raising neuroprotective factors like brain-derived neurotrophic factor (BDNF) and promoting cellular connections. When shared with patients, this neuroplasticity model led to durable changes in self-efficacy and optimism in several studies of depression (Lebowitz MS and Ahn WK, *Behav Res Ther* 2015;71:125–130).

Case Vignette: Brain Chemistry

Jaime, a 33-year-old transgender man with depression, dismissed my exercise recommendation saying, "My depression is chemical—it's about serotonin." At our next visit, I showed him images of neurons before and after treatment. "Exercise and diet cause brain cells to grow and strengthen much as antidepressants do," I explained. Two weeks later, he'd started walking daily. "I feel like I'm actually doing something to fix my brain," he said, "not just waiting for pills to work."

A more direct way to counter passivity is to offer reasonable choices in treatment. Patients respond better in clinical trials when they happen to get randomized to the treatment they preferred from the start. In the shared decision model, clinicians and patients work together to make informed choices about treatment. In this "meeting of the experts," the clinician is the medical expert and the patient is the expert of their life, values, and circumstances. In one model, physicians use laminated cards that highlight the risks, benefits, and costs of various options. David Mintz, who works with chronic depression, actively encourages patients to speak up. "I don't just want you to tell me if I'm doing something that you don't like. I need you to tell me. Otherwise I won't be able to help you as much."

Ambivalence

Most patients have some ambivalence about treatment. They may believe that all good things come with a cost, including recovery. There are side effects to deal with, fears of dependence on medications, and increased expectations from others. As one patient explained after neglecting to start a medication, "I'm afraid that if I get better, my family will unleash all the anger for the problems I caused while depressed. As long as I'm sick, they hold back on it."

Shame

It is difficult to help patients when we don't know what is going on, and this is where shame, stigma, and mistrust get in the way. All clinicians have to weed through dishonesty, but the problem is larger in psychiatry; 70% of patients with depression admit to withholding the truth from their doctor, compared to only 24% with physical illness. The most common subjects where patients with depression withhold the truth are: symptoms, daily activities, medication adherence, and weight (which may explain why physicians are twice as likely to pick up on antipsychotic weight gain with an in-office scale than they are through self-report) (Sawada N et al, *J Clin Psychiatry* 2012;73(3):311–317; Gao K et al, *J Clin Psychopharmacol* 2016;36(6):637–642).

Address the need for openness early, and in a way that conveys that withholding is normal and understandable rather than a problem. "I've heard that patients don't always tell their doctor the truth. While I tell people, 'This is a judgment free zone,' that may not be enough, so I hope you will let me know if I'm doing anything that makes it hard to be open here."

For adherence, leading questions help. Instead of asking "Do you take your medications?" Ask "How many doses do you tend to miss each week?" and normalize it with, "I understand most people don't take their medications as prescribed."

Disability

There are all kinds of secondary gains that reinforce illness, from getting out of responsibilities to disability benefits. Bring this out in the open with a question like, "Is there anything you would stand to lose if you got better?" One patient made great strides in lifestyle changes, but then gave them up, fearing that recovery would cause her to lose her disability benefits. These are difficult choices, and our job is to understand, not to judge. In this case, I clarified that disability is about functioning, not feeling. We reshaped our goal toward living the best life possible under disability, and she moved forward with changes that reduced her suffering.

Transference

Transference distortions are the norm in chronic depression. These patients have learned to distrust authority figures and caregivers like us, as those from their past often ignored their needs or used them to satisfy their own desires. When they reached out for help, they were met with blame and punishment.

The problem may not be obvious at first, as the patient may adopt a position of excessive compliance. They verbally agree with our recommendations and go out of their way to respect our policies. Profuse apologies, excessive deference, and statements like, "I did not want to bother you," are telltale signs. Behind this agreeableness is an expectation that we dislike them and only act caring out of professional obligation or financial reward.

When we disappoint them and their expectations of rejection are fulfilled, compliance can shift to attack. Anger, threats, and accusations follow, and if the damage is not repaired, treatment is disrupted. We become another tale of the callous clinician, much like the one they might have shared about their past provider at their first visit with us.

These disruptions are repairable, and engaging in that work can lead to lasting changes for the patient. Check your pulse, and take a deep breath (this goes better with a cool head). Imagine how the patient expects you to respond, perhaps in some version of attack or defend, and act the opposite. Put aside your own defensiveness. Ask yourself, "How is the patient right," not whether they are right or wrong. Apologize, steer a new course, and thank them for speaking up.

Case Vignette: Mistakes in Practice

Jayden was unfailingly polite in sessions, assuring me it was "completely fine" when I ran late, despite his visible anxiety. When I forgot to call in his prescription before a weekend, he left a calm voicemail reminding me. At the next appointment, he was withdrawn and terse. He eventually acknowledged he was convinced I had "forgotten about me because I wasn't important enough." Rather than defending myself, I apologized and acknowledged his fear: "I'm sorry. I made a mistake. I imagine it made you feel like I had abandoned you when you needed me, just like other people have." This direct acknowledgment, without defensiveness, allowed him to see how childhood patterns were playing out in our relationship and became a turning point in treatment.

Countertransference

Patients who do not get better with our treatments evoke countertransference reactions. We feel disconnected during session, as if we are the only person in the room doing the work for this "help-rejecting complainer." We start to believe they do not want to get better, or doubt our own competence, imagining a specialist whose expertise could get the job done for us.

These feelings often come out in some version of withdrawal or attack. Overt hostility is rare, but attack reactions come out in more subtle forms. Assigning too much homework, refusing requests to adjust medications, and terminating care can all be examples of an attacking countertransference.

Withdrawal reactions happen when we feel hopeless about the patient. We give up on potentially helpful interventions. We daydream during sessions and space out their appointments, convinced that we are saving the patient time and money by seeing them every six months. We forget that our work serves a supportive role even when symptoms don't improve.

Countertransference is difficult to detect because it is unconscious. On the surface, it feels rational and just. The telltale sign is a sense of dread when you see a patient's name on your daily schedule.

Case Vignette: Countertransference Withdrawal

I once treated a patient who complained of intense symptoms at every visit but rejected the suggestions I made. After several months, I realized I was spacing out during our sessions and no longer offered him new ideas. When I recognized this pattern, I sought consultation and discovered I had fallen into withdrawing from a patient who made me feel ineffective. At our next appointment, I acknowledged the pattern: "I notice I've stopped bringing new ideas to our sessions, and I think that's because I'm afraid they'll go nowhere or get rejected." This opened a conversation about how his defensive self-sufficiency was keeping potential allies at a distance, allowing both of us to re-engage with treatment.

The Real Placebo

The placebo is sometimes called a sugar pill, but the two are not the same. The placebo effect includes the natural course of illness, the therapeutic relationship, supportive factors in the patient's life, the patient's resilience, and expectations about treatment.

Patients are more likely to respond to medications they perceive as powerful, such as those that are brand name, expensive, or come in certain colors (red is associated with stimulation, blue with sedation, and yellow with happiness). Pharmaceutical companies strive for unique names with "Zs" and "Xs" in them, but they can't push it too far. The effect of naming is so powerful that the FDA forbids drug names that overstate their promise. That is why varenicline is called "Chantix" in the US and "Champix" outside the US (somehow "Abilify" and "Adderall"—as in "ADD for All"—slipped past the auditors).

Patients with Parkinson's disease also respond to a placebo. Patients who are immobilized by Parkinson's disease are suddenly able to rise and walk when shown a novel pattern of gold and black squares on the floor. Why? Perhaps the novel imagery triggers dopamine release. The problem is that the novelty wears off. These flash-in-the-pan responses are not what we need in chronic disorders. Eventually, they lead to disappointment and polypharmacy. Instead, we need to rely on the parts of the placebo that support more lasting change.

A strong therapeutic alliance is one. This develops when patients experience us as trustworthy, consistent, genuine, warm, and empathic. It is further enhanced when we seek to understand the patient as a person, find something we like in them, and take a nonjudgmental attitude toward all that they bring. But it can also rupture. Patients often wonder if our empathic, professional stance is genuine or just a show. They watch how we interact with office staff, and how we act in chance encounters outside the office.

The power of the placebo is generally greater than that of an antidepressant. In trials of depression, 75% of the response is due to placebo, and a trial sponsored by the National Institute of Mental Health (NIMH) shows us how much that effect depends on the therapeutic alliance. In the 1980s, the NIMH conducted a large trial to compare an antidepressant (imipramine) with two forms of psychotherapy (interpersonal and cognitive behavioral therapy). In the end, all had similar benefits, but when the researchers reanalyzed the antidepressant and placebo groups 20 years later a different pattern emerged.

In this trial, the patient's chance of response depended more on which psychiatrist they saw than whether they got the placebo or the antidepressant. Put another way, among the nine psychiatrists who took part, the top 30% got better outcomes with placebo than the bottom 30% did with the antidepressant (McKay KM et al, *J Affect Disord* 2006;92(2–3):287–290).

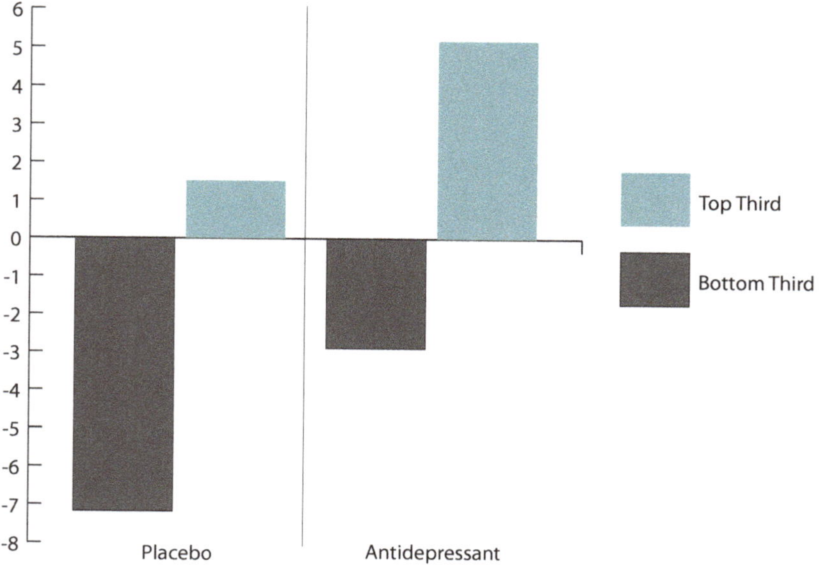

Psychiatrists from the NIMH Collaborative Depression Trial were divided into the top and bottom third based on their subject's outcomes. In the graph, psychiatrists from the top third achieved better responses with placebo than those from the bottom third did with imipramine.

FIGURE 12-3. Antidepressant and Placebo Response Varies by Psychiatrist

A strong therapeutic alliance inspires hope and improves medication adherence. Like a safe harbor, it provides security, allowing patients to take on risks and try new things. In the next chapter, we'll look at how to use the alliance to help patients take an active role in their recovery.

Key Takeaways

- The therapeutic relationship contributes more to the outcome than specific medication effects—cultivate genuineness, warmth, and empathy.
- Depressive symptoms themselves—hopelessness, passivity, worthlessness—get in the way of treatment.
- Patients with chronic depression have high rates of childhood adversity, and this affects their trust in caregivers.
- Watch for transference distortions and respond with efforts to repair. Avoid the common countertransference traps of defensiveness, anger, and withdrawal.

- Use the neuroplasticity model to show patients how lifestyle changes enhance medications through common pathways, and engage them by offering reasonable treatment choices.

CHAPTER 13

Working with Chronic Depression

AT SOME POINT IN THE TREATMENT of difficult depression, we face a pivotal question: What happens when medication trials offer diminishing returns? This chapter addresses that critical juncture where our focus must shift from pursuing complete remission to helping patients build meaningful lives despite persistent symptoms.

The chance of remission goes down with each successive antidepressant trial (Petersen T et al, *J Clin Psychopharmacol* 2005;25(4):336–341). This doesn't mean we need to abandon medication altogether. Some patients will respond to specialized approaches like high-dose monoamine oxidase inhibitors (MAOIs), pramipexole (Mirapex), transcranial magnetic stimulation (TMS), or electroconvulsive therapy (ECT). However, it does signal a necessary shift in our therapeutic approach.

Patients with chronic depression often feel trapped in an endless cycle of medication trials, each promising a cure that never materializes. Recognizing this turning point isn't about giving up—it's about adopting a more realistic approach.

Reframing the Therapeutic Goal

Joe Goldberg, who specializes in treatment-resistant depression, articulates this shift clearly: "We shouldn't equate pharmacologic futility with hopelessness about the ability to manage chronic disorders. We have to maintain some sense with the patient of 'We'll get you through this,' though it may not be with pills."

This reframing requires candid communication. He often tells patients, "I hope we'll find something that's transformative, but this depression has been with you a long time. The chance that it will completely resolve in the near future is low. Let's talk about how you can live your best life possible while managing these symptoms."

This isn't about lowering expectations—it's about redirecting energy toward achievable goals. The conversation shifts from "I've got a thorn in my side, and I need you to take it out" to "You've got a thorn in your side that's not moving, so how are you going to go to work with that thorn in your side?"

Mobilizing Patient Resources

One of the greatest challenges with chronic depression is the passivity it encourages. Patients become defined by what they can't do rather than what they can. Our job is to help them rediscover their agency.

When patients return for a follow-up, Dr. Goldberg focuses on two domains: symptoms and functioning. "How are you feeling, and what are you doing? If they report feeling slightly better but remain inactive, I express genuine puzzlement: 'Your mood is improving, but you're still in bed most of the day. Why is that?'" This gentle confrontation highlights the disconnect between symptoms and behavior.

Depression doesn't create many positive emotions to mobilize patients, but it often results in anger. When properly directed, this anger can become fuel for change. Some chronically depressed patients benefit from appeals to their better self. As Goldberg explains, "I'll try to mobilize the patient against the illness, and I may appeal to their narcissism there. 'You're better than this. The illness is the enemy, and it's getting in the way of your ability to tap into your inner reserves.'"

The Paradoxical Approach

Some patients with chronic depression present as "help-rejecting complainers." They seem to fight against our interventions while demanding relief, leaving clinicians feeling ineffective and demoralized.

For these patients, a paradoxical approach can help. Instead of pushing treatment they'll resist, we temporarily align with their pessimism.

The dialogue might unfold like this:

Clinician: "You have a tough depression. I'm not sure I have anything that can work, and there's a good chance that you'll have a lot of side effects if we start something new."

At this point, patients often begin pushing back, insisting on trying something and wanting to prove you wrong. After starting the suggested treatment:

Patient: "I think I'm feeling a little better."

Clinician: "Well, it's a little too soon for the medication to be working, so I have to warn you that it may not last."

Patient: "No, I'm optimistic. I think we're on to something."

Clinician: "Yeah, but I bet you're having a lot of side effects."

Patient: "No, I don't have any side effects!"

This approach works by converting the patient's resistance into motivation. Instead of fighting against treatment, they're now fighting against the notion that they can't improve. The key is genuine empathy—this technique fails if it comes across as manipulative or dismissive.

Focus on Functioning

When medication approaches reach diminishing returns, functional improvement becomes our primary goal. This means helping patients increase activity and engagement even while symptoms persist.

Effective strategies include:

- **Simplifying**—Behavioral activation sometimes fails because it doesn't account for the severe fatigue and anhedonia of depression. Simplify the assignments, and schedule time to relax and reflect on the accomplishment (no matter how small).
- **Energy budgeting**—Help patients identify their highest-value activities and allocate their limited energy accordingly. This might mean skipping housework to attend a child's soccer game or taking a taxi instead of the bus to conserve energy for a job interview.
- **Planning for symptom interference**—For each major symptom, develop specific workarounds. For example, if concentration problems make reading difficult, audiobooks might be an alternative.
- **Celebrating functional wins**—When symptoms are chronic, even small functional improvements deserve recognition. These successes build momentum toward larger changes.

The Role of Hope

Throughout this process, we must maintain hope—not for a miraculous cure, but for a meaningful life despite symptoms. As Goldberg notes, "The message is, 'You can work with this. People with chronic illness—whether spinal cord injuries or depression—have a way of mobilizing their resilience. I don't know how that's going to look for you, but I know that you are capable of it.'"

This realistic hope acknowledges difficulty while affirming possibility. It says, "This is hard, and you may always have to manage these symptoms, but you can still create a life worth living."

Key Takeaways

- When symptoms are unremitting, focus on functioning.
- Engage patients as active managers of their condition rather than passive recipients of treatment.
- Anger is an understandable reaction to chronic illness, and patients can mobilize that energy to build a more meaningful life.

CHAPTER 14

Therapy and Lifestyle

"HERE'S WHAT DEPRESSION IS LIKE," explained a patient. "All day long there's a movie playing in my head of the things I should be doing, and I don't do them, and I don't know why I don't do them." It is a vivid description of what happens when the connections between the action center (prefrontal cortex) and the emotional center (amygdala) break down, as they do in depression. She knows what to do but doesn't have the emotional thrust to get up and do it.

Other patients simply say they are "overwhelmed." They are unable to prioritize the long list of tasks that shuffle through their mind, or separate that list from the negative thoughts and emotions that scurry around it. Our job is to help them prioritize by highlighting lifestyle changes that are simple enough to take on but powerful enough to make a difference in depression. However, before we add anything, they might need to remove a lot from their task list.

Reducing Expectations

While we don't want to encourage total passivity, we also don't want to overwhelm patients by putting too much on their plate. Gerald Klerman and Myrna Weissman struck a helpful balance in this dilemma when they developed interpersonal psychotherapy (IPT) in the 1970s. Borrowing from the sociologic concept of the "sick role," they explain to the patient that their depression makes them unable to fulfill their usual responsibilities. However, this isn't a free pass. Much like a person who has tuberculosis, they are expected to seek treatment and work toward recovery. When delivered well, the result is less self-flagellation and more focus on therapeutic goals.

In IPT, those therapeutic goals involve working to resolve a recent stress, such as a relationship conflict or a major life change. The therapy works great for acute depressive episodes, but it has a limitation that may explain why it didn't work in chronic depression (Markowitz JC and Weissman

MM, *Clin Psychol Psychother* 2012;19(2):99–105). IPT requires patients to understand other people and take a logical approach to solving problems. That might work for someone who functioned well before the depressive episode and has a healthy baseline to draw from, but those skills are in short supply in chronic depression.

Compared to those with acute episodes, patients with chronic depression are more likely to have complex problems and less likely to have the tools to solve them. They have higher rates of personality, anxiety, and substance use disorders and function at a lower level in work and relationships. We don't want to demoralize them further with unrealistic goals that are a set-up for failure. In this chapter, I've selected goals that don't require as much social skill but still have a meaningful effect on depression.

TABLE 14-1. Lifestyle and Depression: A Patient Guide

Movement	Brisk walking 30–45 minutes every other day
People	Too much interaction is stressful, but a minimum amount of supportive contact gives the brain the sparks it needs to make antidepressants work.
Activity	Look for activities that are absorbing or in line with your goals or values. Start by adding a few good minutes of them to each day.
Rhythm	Set your biological clock with morning sunlight and evening darkness. Wake up at regular times, stay out of bed during the day, and schedule a wind-down time before bed.
Sleep	If you have insomnia, there are behavioral steps that improve mood and sleep.
Nature	Walking or just being in nature two hours a week reduces depression and negative cycles of worry.
Food	A Mediterranean-style diet rich in fruits, vegetables, nuts, beans, fish, whole grains, and lean meats has a big effect on depression.
Health	Mental health improves when you take better care of your physical health. This is true for nearly all health problems, from diabetes to dental health, smoking to sleep apnea.

1. Movement

Aerobic exercise treats depression as effectively as an antidepressant, but the word "exercise" is best avoided. Instead, use "movement" or "walking."

Aerobics means anything that gets you breathing faster and raises your heart rate by 10 bpm.

Brisk walking accomplishes this, and the dose for depression is 30–45 minutes every other day (more strenuous levels may be needed for depression with inflammation). Start smaller, with 3–5 minutes a day and build from there. Use a timer to gamify the movement, or have the patient move to the beat of inspiring music, starting with the duration of a single song.

This kind of exercise augments antidepressants, works in treatment-resistant depression, and prevents depression better than a selective serotonin reuptake inhibitor (SSRI) (Babyak M et al, *Psychosom Med* 2000;62(5): 633–638). It raises neuroprotective factors like brain-derived neurotrophic factor (BDNF) and has additional benefits that antidepressants lack, such as in cognition, sleep, and physical health. That is why the National Institute for Health and Care Excellence (NICE) guidelines in the UK recommend exercise or psychotherapy instead of an antidepressant for mild depression.

2. Absorbing Activity

People with depression long to get out of their head, to turn off the nagging cycle of negative thoughts. One of the best cures for rumination is absorbing activity. We're looking here for things that have a stronger gravitational pull than the rumination that otherwise absorbs them. Activities where it's hard to think about anything else while doing them, like riding a bike or playing a game.

Ask, "Is there anything you do that gets you out of your head, where time flies and you're not as self-aware and self-conscious? Did you have good moments in the past week? What were you doing then?" Here are some qualities that absorbing activities tend to share:

- **Challenging enough**—It's difficult enough to keep their attention but not so challenging as to be overwhelming.
- **Clear goals, instant feedback, and sensory involvement**—It's easier for people to keep focused when they know what they're aiming for and can see where they're going. In painting, every stroke changes the picture; when cleaning, every motion removes a little more dirt. When in doubt, ask what they enjoy doing with their hands. Cooking, gardening, sports, musical instruments, photography, and crafts are good examples. Videogames also fit the bill, but if the activity is too absorbing (ie, addictive), limit it to 30 minutes a day.

- **Time flies**—People lose their sense of time during absorbing activity. They don't think about the future or the past, or watch the clock, wondering when it will be over.
- **Lack of self-consciousness**—Their focus is on the outside world, other people, or a greater cause, rather than on themselves. If someone interrupts to ask how they're feeling, they might not even know the answer.
- **A higher cause**—Behavior change is difficult to sustain in depression. Few things are rewarding enough to reach the tracks of reinforcement in the depressed brain. That trigger is more likely if the activity is in line with their values or goals. Look for activities that serve a purpose beyond their own needs and desires. Things they do for the love of the game, regardless of the outcome. Volunteering at an animal shelter, playing games with their children, learning, creating, or reading a religious or spiritual text.

After your patient chooses an absorbing activity to add to their day, ask them to decide on a time and place when they will do it. Behavior change rarely happens without those two details. Next, troubleshoot in advance by asking about common obstacles that might get in the way. If the activity is playing basketball, what if it rains? What if they feel tired? Or forget? How will they remember to keep their commitment? In the end, remind them that there is no such thing as failure in this work. If they don't take action, it's an opportunity to problem-solve, learn, and plan a different course.

With regular application, this technique makes a big difference. It is derived from behavioral activation, one of the few therapies that works in severe depression (Dimidjian S et al, *J Consult Clin Psychol* 2006;74(4):658–670). Behavioral activation quiets the area of the brain responsible for depressive rumination, the default mode network (Yokoyama S et al, *J Affect Disord* 2018;227:156–163). It is part of cognitive behavioral therapy (CBT) and may be the most effective part. In a large study that compared six components of CBT, absorbing activity led to the most significant and lasting difference in depression (Watkins E et al, *JAMA Psychiatry* 2023;80(9):942–951).

3. Rhythm

Depression is caused in part by a broken biological clock. Night owls are more likely to develop depression, and when they correct that sleep phase-delay their mood improves (Wescott DL et al, *Sleep Med Rev*

2025;79:102022). Working the night shift raises the odds of depression by 22%, even after adjusting for other depressing confounders that go along with shift work (Xu M et al, *JAMA Netw Open* 2023;6(8):e2328798).

Social Rhythm Therapy treats mood disorders by regulating circadian rhythms. First developed for bipolar disorder, it has since shown a large effect in unipolar depression (Yau AY et al, *J Consult Clin Psychol* 2024;92(3):135–149). Start with the bookends of the day, morning and night.

Waking up at a regular time* each day stabilizes sleep and improves mood, particularly if they wake up to sunlight. Unfortunately, a little-known symptom of depression can get in the way here: sleep inertia. This means the brain is slow to wake up, making people feel tired, sluggish, and foggy in the morning. Sleep inertia is common outside of depression, but for most people it only lasts about 15 minutes. In depression, it can go on for hours (before diagnosing it, make sure to rule out medication side effects, including serotonergic antidepressants, which can disrupt sleep).

Dawn simulators reduce sleep inertia by creating a gradual sunrise in the bedroom. The light rises over 30–60 minutes, gently lifting people from deep sleep to full awakening. In contrast, sound alarms jolt people out of deep sleep, disrupting sleep architecture and leaving them in a hazy state of sleep inertia. On average, a dawn simulator cuts the duration of sleep inertia in half (modafinil 200 mg has a similar benefit). Dawn simulators also treat seasonal winter depression, but are probably not as powerful as a lightbox.

When using a dawn simulator, the patient should wake up gradually, about 30 minutes after the light begins. If they wake up too soon, move it away from the bed. If they don't wake up, move it closer. Research-backed models cost $100–$120 and include Philips SmartSleep and Lumie Body Clock 100. Budget models are about half the cost and include Winshine Touch, JALL Wake Up Light Sunrise Alarm Clock, and the Philips Hue lightbulbs (which can be programmed to turn on gradually through a smartphone).

To stabilize nocturnal rhythms, establish a wind-down routine at a regular time before bed. Patients can find what works best, but typical ingredients include dim lights, relaxing music, and simple, calming activity in an electronic-free zone. If they are a night owl or have trouble sleeping, blue

* In circadian research, "regular" means give or take 15 minutes.

light blockers can help. These amber glasses block the blue wavelengths of light that suppress melatonin. Wearing them one to two hours before bed will help patients fall asleep a little more easily, but they have a bigger effect on deepening sleep. The result is better functioning the next day.

To achieve this effect, they'll need a pair of glasses that blocks close to 100% of blue light. They'll also need to sleep in a pitch-dark room when they take the glasses off to go to bed, such as with blackout curtains. Most glasses don't block enough blue light. Those that do include the budget Uvex Skyper S1933X ($10–$20) or any pair at LowBlueLights.com or CET.org ($40–$60). Those sites have larger "fit over" models for patients who wear glasses. If your patient is unable to sleep in the dark, they can search for a blue light-free nightlight on those websites or on Amazon.

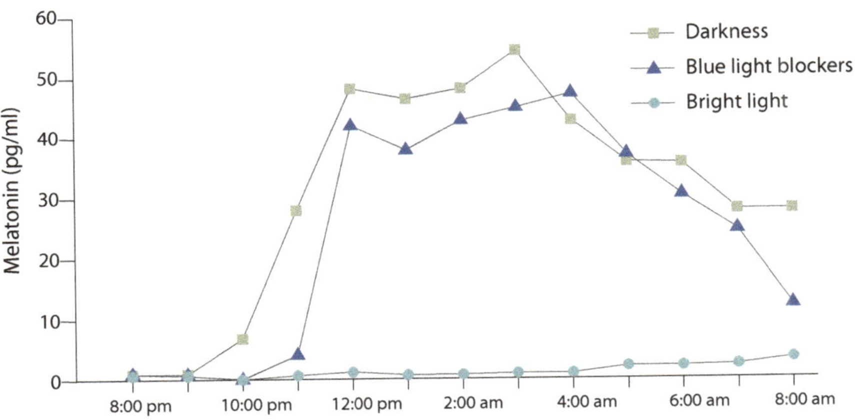

FIGURE 14-1. Evening light suppresses melatonin, but melatonin levels rise when people are in a dark room or wear blue light blocking glasses

Source: Kayumov L et al, *J Clin Endocrinol Metab* 2005;90(5):2755-2261

4. Sleep

In 2013, *The New York Times* announced "the most significant [advance] in the treatment of depression since the introduction of Prozac." The discovery was a behavioral therapy for insomnia that had an unexpected side effect: It improved depression. The treatment is cognitive behavioral therapy for insomnia (CBT-insomnia), and its ability to lift mood has held up in more than 40 controlled trials of depression and insomnia.

CBT-insomnia is different from sleep hygiene, a more basic approach that guides patients to wake up at regular times and avoid daytime naps, afternoon caffeine, and trying to force sleep by going to bed before getting tired. This is all good advice for an insomniac, but it doesn't restore sleep or improve mood as well as CBT-insomnia (sleep hygiene is often used as the inactive control in studies of CBT-insomnia).

The distinctive ingredient in CBT-insomnia is bed restriction. We induce a mild sleep deprivation by limiting the patient's time in bed. This raises their sleep drive so that—eventually—they fall asleep more naturally. We calculate the amount of time allowed in bed with this formula:

Average time spent asleep in bed per day (including naps)
plus a 30-minute buffer

However, it's best not to restrict the time in bed to less than 5.5 hours, so if the number comes out low, round up to 5.5. In the formal therapy, the patient calculates this average by recording their sleep times every night over a week. In practice, you can make a big difference with a rough-and-ready approximation by using 5.5 hours or asking the patient how long they spent asleep each day and adding 30 minutes.

Next, figure out a workable start time to get in bed. For example, if their bed allowance is 6.5 hours and they have to wake up at 8:00 am each day, they should get in bed at 1:30 am. Other patients may choose a time when they are most likely to fall asleep, say 11:00 pm, in which case their wake time would need to be 6.5 hours later (5:30 am).

The difficult part is inspiring the patient's trust in this work, and here a little knowledge of sleep physiology helps. "I understand sleep medications haven't worked for you, but that's not a surprise. Sleep medicines are weak. They just haven't figured out a chemical that will do all the complex things that sleep does, allowing you to fall asleep while still breathing and not acting out your dreams. However, your body has two natural sleep medicines that work better: melatonin and adenosine."

"Melatonin controls your circadian rhythm. It rises when the lights are out and falls when the sun comes up. You can regulate melatonin by keeping a regular wake time and getting your room pitch dark at night. Adenosine controls your sleep drive. It rises throughout the day, so that the longer you're awake the more there is. In this therapy, you're going to maximize adenosine through controlled sleep deprivation. You'll also need to avoid caffeine. That suppresses adenosine, just as light at night suppresses melatonin. After a few weeks these two chemicals will start to align and you'll fall asleep more naturally."

As sleep improves, patients can increase the time allowed in bed (or restrict it further if sleep worsens again). CBT-insomnia is often delivered in a self-guided format, such as through prescription apps (SleepioRx, Somryst), a free app (CBT-I Coach), or online seminars and books. Those formats work almost as well as face-to-face sessions, even for patients with depression (Ho FY et al, *J Affect Disord* 2020;265:287–304).

5. Nature

Studies of greenspace and health are plentiful, inspiring the unofficial term "Nature Deficit Disorder." People seem to need a minimum effective dose of nature to sustain well-being. While the amount probably varies by person, studies suggest that two to three hours a week is a good starting place. Time spent in the forest or around bodies of water works best.

A 90-minute walk in the woods improves depressive rumination more than a similar walk in the city or suburbs. That nature walk also turns down the default mode network that drives that spiral of worry and self-critical thought (Bratman GN et al, *Proc Natl Acad Sci U S A* 2015;112(28):8567–8572). Nature walks improve ADHD and are associated with improvements in blood pressure, cortisol, and immune function (Qiu Q et al, *Int J Environ Res Public Health* 2022;20(1):458). Some of those biological changes are measurable for up to a week after the walk.

The possible mechanisms here are many. Plants emit phytoncides in the air, organic compounds that improve immunity. Outdoor air is rich in sunlight and negative ions, both of which have known antidepressant effects. On a psychological level, nature has a way of engaging our attention without becoming boring or requiring too much effort. The result is a mindful mental state called "soft fascination." You might find it by watching clouds form in the sky or following the grains in a wood-paneled wall.

6. Food

Food affects the brain much as it does the other organs. A healthy diet supplies the brain with the building blocks for neuronal membranes and neurotransmitters like omega-3 fatty acids and folate. When patients shift to brain-friendly foods, neuroprotective factors rise, inflammation falls, and stress hormones stabilize. In epidemiologic studies, Western-style diets raise the odds of depression 18%, while healthy diets lower it by 36% (Li Y et al, *Psychiatry Res* 2017;253:373–382).

Researchers have known this for years, but when Felice Jacka and colleagues at Deakin University in Australia launched the first randomized controlled trial of a dietary intervention for depression, they were surprised by the magnitude of the results. Dietary counseling treated depression with a large effect size (1.2) compared to a control group who were given a treatment with benefits of its own, supportive therapy. The antidepressant effects did not correlate with changes in weight, activity, or a sense of self-efficacy, but with the food itself. For every 10% change in diet, there was a 5% drop in depression scores (Jacka FN et al, *BMC Med* 2017;15(1):23).

Independent trials confirmed those results, even as they decreased the intensity of the dietary counseling, from hour-long, weekly individual sessions to group seminars to a single 13-minute video. That makes this an apt intervention for a brief therapy visit (Parletta N et al, *Nutr Neurosci* 2019;22(7):474–487; Francis HM et al, *PLoS One* 2019;14(10):e0222768; Bayes J et al, *Am J Clin Nutr* 2022;116(2):572–580).

The diet employed in these studies is a Mediterranean style approach. There is no calorie counting. The diet was designed to be easy to follow, so you don't need to be a nutritionist to teach it. Nor do you want your patient to stress over perfection in this work; enjoying the food is one of the ingredients.

The table divides the diet into foods to:

- **Enjoy in abundance**—These do a lot of good with little harm.
- **Appreciate in moderation**—Foods that deliver brain-essential ingredients (eg, choline and PEA from eggs, iron and B-vitamins from meat) but have risks in high doses.
- **Eat less**—Foods that mostly cause harm.

Start by asking patients to find foods they enjoy in the first category (foods to enjoy in abundance). As their intake of those increases, the less desirable foods will naturally decrease. Encourage patients to shift their focus from how the food tastes on their tongue to how it makes them feel a few hours later. Does it make them more foggy-headed and tired, or sharper and motivated? Ask them to keep a food journal and share the results with you.

Strict adherence is not necessary. Patients who followed half the diet still got half the benefits in the clinical trials. For patients who can't subscribe

TABLE 14-2. Mediterranean Diet for Depression

Enjoy in Abundance	Fruit	Fresh or frozen are fine; dried and juiced fruits are a bit sugar heavy
	Vegetables	Aim for a variety of colors; olives and mushrooms count
	Beans	Hummus counts
	Fish	Especially oily fish, like salmon, tuna, sardines, anchovies, and mussels
	Nuts and seeds	Salt-free is best; nut butters with low sugar and minimal processing are great
	Whole grains	Brown rice, oatmeal, quinoa, homemade popcorn, and breads or pasta with "100% whole grain" on the label
	Extra virgin olive oil	The "extra" means more antioxidants, and it's safe to cook with up to 375°F
Appreciate in Moderation	Chicken and lean red meats	One serving/day of either meat; avoid fried chicken, bacon, and processed deli meats
	Eggs	Six eggs/week
	Dairy	One serving/day; avoid processed cheeses like Velveeta and yogurt with artificial flavorings.
Eat Less	Fried & fast foods Processed foods Sodas Sweets Simple carbs Refined flour Bacon & sausage Deli meats Condiments	Limit to three small servings a week of any of these foods (small = 120 calories) Choose packaged foods with low sugar, salt, and fewer processed chemicals; the Open Food Facts App rates packaged foods by scanning the bar codes
Drinks	Black and green tea lowers depression risk (up to 6 cups/day) Coffee limit to 2–3 8 oz cups of regular or 2–4 shots espresso/day Avoid alcohol or limit to 1 drink every other day (prefer red wine)	Aim for drinks low in sugar, salt, and chemical ingredients

to the whole diet, some of the ingredients can make a big difference with a small change. Eating a cup of blueberries or a half cup of almonds each day improved depression and cognition in several randomized trials (the almond study involved patients with depression and type II diabetes) (Aiken C, *Psychiatric Times*, February 9 and May 19, 2021).

7. Health

Self-neglect is a symptom of depression. That means fewer showers, lower medication adherence, and skipping out on primary care visits. Most medical illnesses worsen mental health, in particular those that cause inflammation or affect the vascular, endocrine, metabolic, sleep, or neurologic systems.

Poor dental health also has causative links to depression, and not just because it reduces quality of life. The *Porphyromonas gingivalis* pathogen responsible for gingivitis causes inflammation and reduces neuroprotective factors in the brain (Wang YX et al, *Brain Behav Immun* 2019;81:523–534).

Behavioral strategies can help patients make changes here, for example:

- **Medication adherence**—Have them link their medication to something they already do each day, such as morning coffee. For multiple medications, a pill box is essential (they can buy several to reduce the chore of filling them up).
- **Appointments**—Scheduling medical appointments is not easy. Start by having them plan a time and place when they will make the phone call, and put it on their calendar.
- **Dental care**—Dentists recommend brushing two to three minutes twice a day. The task is easier when done to a favorite tune, and most popular songs last about that long.

8. People

Social isolation is one of the top predictors of antidepressant nonresponse. In animal models of depression, antidepressants don't work as well when the animals are kept in isolation (Rief W et al, *Neurosci Biobehav Rev* 2016;60:51–64). Patients with depression often find social interactions aversive, but they probably need a minimum level of social engagement to keep their neurons operational.

As important as this item is, it's also one of the more difficult to take on, so I've put it last. I wouldn't recommend it unless the patient is particularly motivated to improve their social life, and even then, I would wait until they can meet weekly with a therapist to troubleshoot problems that arise. Patients who are isolated or have poor social skills should start by socializing in a structured environment like a volleyball team, a gardening club, or a church choir. When groups have clear rules and expectations, patients are less likely to interpret interactions as attacking or rejecting.

Putting it into Practice

When a patient starts a new medication, I'll write two prescriptions. One for the medication and one for a lifestyle change. "Medication works like a spark to a fire, but fire needs oxygen and dry wood. Likewise, your brain needs to have a few things in place for the medication to work. You don't need a lot; just the bare minimum." Then I'll ask them to choose one from the eight options in Table 14.1 (Lifestyle and Depression: A Patient Guide).

Most patients have multiple areas they could work on, so where to start? The best choice is the one they are likely to succeed with, so start with a single change, collaborating with the patient to find one that is:

- Simple, easy, and a good match for their strengths
- Something they care about and are motivated to take on
- Suitable for the type of therapy you can provide

Change requires troubleshooting. The more complicated the change, the longer and more frequent the sessions they will need. The work involves translating the goals into specific, measurable changes. "I need to take care of my diabetes" becomes "At 2:00 pm on Tuesday I will call my endocrinologist to schedule an appointment." It helps to have them record their progress so that they'll have something concrete to share with you at the next session.

Lack of progress is common. Respond positively, with interest and curiosity, about what got in the way. The table lists four common reasons why patients don't follow through with behavioral goals they set. The first three are relatively straightforward. The last one, experiential avoidance, is trickier and is also the most common. Patients don't make changes because doing so would bring up uncomfortable feelings.

This experiential avoidance is widespread in depression, and it even comes up when asking them what got in the way of making the change.

When patients respond by shutting down or giving vague answers like "I don't know," they are enacting that avoidance in the room, and avoidance is likely the thing that got in the way. You can shift to an easier task or look for one that has a stronger connection to their values.

When patients are unable to commit to any change, ask them to schedule an hour each day to do nothing. Presumably, that is what they are already doing, but they are doing so with a sense of passivity and a heavy feeling of guilt about their inaction. Scheduling it moves them closer to intention and acceptance.

If the assignments keep running into avoidance, they will likely need more dedicated sessions, such as meeting for 45–60 minutes a week. In that work, the therapist responds to in-session avoidance by empathically moving toward it. "What are you feeling right now? I see that you are struggling, and I'd like to ask you to gently move into the experience rather than fighting it, and, without judgment, describe what you are feeling. Where do you feel it? I'm here with you and am open to anything you may experience."

TABLE 14-3. Reasons for Inaction

Reason	Solution
Forgot	Have them set a calendar alert on their smartphone, a sticky note in a prominent place, or involve a supportive person to remind them.
Skill deficit	For social skills (eg, meeting people), consider a more dedicated psychotherapy. For nonsocial skills (eg, how to cook a fish or find a dentist), consider courses, online learning, and practice.
External factors	Here the change would cause problems with other people or conflict with other responsibilities. Solutions include bringing the other person into session or problem solving.
Experiential avoidance	Here the change causes internal sensations that the patient wants to avoid (usually anxiety). Solutions include psychotherapy with a focus on distress tolerance and connecting the changes to the patient's values.

Key Takeaways

- Lifestyle interventions enhance antidepressants and provide the necessary neural environment for medications to work.
- Start with a single, manageable change that the patient is motivated to adopt rather than overwhelming them with multiple recommendations.

- The most common barrier to behavioral change is experiential avoidance—the desire to avoid uncomfortable feelings that arise when patients start to take an active role in their lives.

Biological Treatment

CHAPTER 15

When Antidepressants Don't Work

WHEN A PATIENT EMBARKS ON their first antidepressant, their chance of responding follows the rule of 3s:

- 1/3 reach remission
- 1/3 respond partially (at least a 50% improvement)
- 1/3 respond minimally or not at all

Those are rough approximations, as the actual rates vary by population (Al-Harbi KS, *Patient Prefer Adherence* 2012;6:369–388). When patients don't have a meaningful recovery, the usual approach is to switch to a new antidepressant or augment with another medication.

Should You Switch or Augment?

After the first antidepressant fails, most clinicians switch to a different class. This often works in practice, but in most cases, it is the placebo that is doing the work. Switching was no more effective than staying with the original antidepressant in an analysis of eight randomized trials (Bschor T et al, *J Clin Psychiatry* 2018;79(1):16r10749).

Augmentation, however, is twice as effective as placebo after the first antidepressant fails (Zhou X et al, *J Clin Psychiatry* 2015;76(4):e487–e498). This is true regardless of whether a patient had a small response or no response to that antidepressant. However, many augmentation strategies carry risks that make them more suitable when the problem is severe, like lithium, thyroid, and the antipsychotics.

Switching antidepressants may not work as well as augmenting for the average patient, but it may be the right move for some depressive subtypes. Monoamine oxidase inhibitors (MAOIs) are preferred for atypical features, and tricyclics for melancholic features.

If you do decide to switch, you may need to give that second trial more time to take effect. While a standard antidepressant trial is six weeks, more time may be needed—up to 12 weeks—to assess response to a second trial (see figure) (Bschor T et al, 2018).

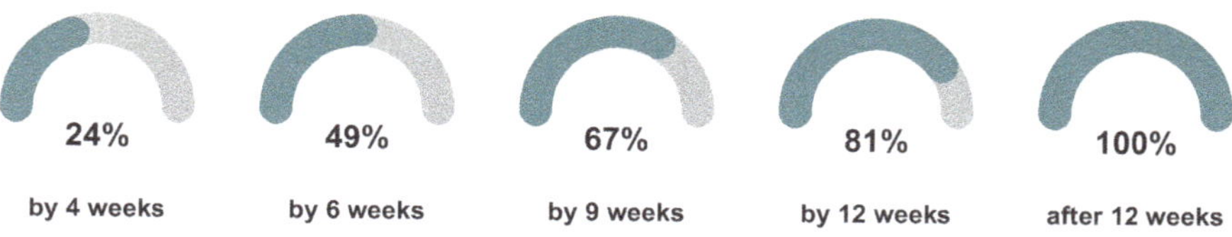

FIGURE 15-1. Chance of Response to a Second Antidepressant Trial

Optimizing the Dose

Raising the dose usually does not work unless your patient is still taking a starting dose. Most antidepressants have a flat dose-response curve, which means that response levels off beyond the optimal dose ranges, which are listed in the table.* Side effects, however, continue to rise as the dose goes up. The main exception to this rule is the tricyclics, which become more effective at higher doses, and many tricyclics have well-defined serum levels that can guide the dose. Other possible exceptions include desvenlafaxine (Pristiq) and the MAOIs (and, outside the US, reboxetine), all of which tend to have stronger antidepressant effects as the doses increase (Hamza T et al, *Stat Methods Med Res* 2024;33(8):1461–1472).

How then do we explain patients who respond to higher doses of antidepressants? For many, it is a placebo effect, which includes the natural course of illness and the fact that they might have gotten better if given more time. But there are probably a few—around 5%—who are off the bell curve of these studies and benefit from a higher dose.

Effective Augmentation Strategies

Among pharmacologic strategies, only four have clear evidence to augment antidepressants: lithium, second-generation antipsychotics, pramipexole,

---

*Of the approximately 39 meta-analyses that have examined dose-response curves with modern antidepressants, the majority (80%) found no improvement with higher doses. The few analyses that found the opposite included more biased trials in their analyses, such as flexibly dosed trials, which insert more uncontrolled variables than fixed-dose trials.

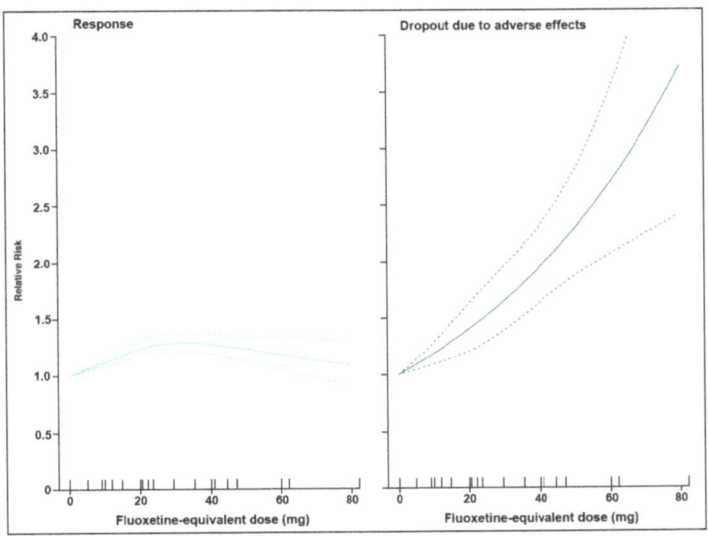

FIGURE 15-2. Response and Side Effects by SSRI Dose

[Dose-response curves for response and side effects to SSRIs in depression, based on 99 fixed-dose groups. Source: Furukawa TA et al. *Lancet Psychiatry* 2019;6(7):601–609.]

and the ketamines. Among those, lithium has the best preventative effects and intravenous ketamine has the largest acute effects.

Promising But Less Certain Strategies

Many popular augmentation strategies have come up short in controlled trials, like bupropion (Wellbutrin), buspirone (Buspar), mirtazapine (Remeron), methylphenidate (Ritalin, Concerta, etc), and the amphetamines (Adderall, Vyvanse, etc). Often they showed promise in early trials, only to fail in larger studies that have better designs. Unfortunately, by that time, their adoption has become widespread, and the failure in larger trials does little to curtail their use. At the end of this book, we'll look at where these popular strategies may still have a role and when they should be tapered off.

Between the proven augmentation strategies and the disproven ones are a host of second- and third-line options that are supported by small randomized trials but have not yet been tested in large ones. These include thyroid, tricyclics, novel medications—amantadine, celecoxib (Celebrex), d-cycloserine, and minocycline (Minocin)—and some complementary and alternative treatments.

TABLE 15-1. Optimal Dose Ranges for Second-Generation Antidepressants

Antidepressant	Optimal Dose (mg/d)*	FDA Max in Depression
Citalopram	20–40	60
Escitalopram	10–20	20
Fluoxetine	20–40	80
Fluvoxamine	100–150	n/a
Paroxetine	20–30	50
Sertraline	50–100	200
Duloxetine	40–60	120
Desvenlafaxine	50–200	400
Venlafaxine	75–150	225
Bupropion	150–300	450
Mirtazapine	15–30	45
Nefazodone*	300–600	600
Trazodone*	150–300	600
Vilazodone	10–20	20
Vortioxetine	20–40	40

*Optimal doses are derived from dose-response curves from fixed-dose studies; that data is not available for nefazodone and trazodone (Furukawa TA et al, 2019; Hamza T et al, 2024)

Nonpharmacologic Approaches

Both electroconvulsive therapy (ECT) and transcranial magnetic stimulation (TMS) are more effective than pharmacologic augmentation. Light therapy also works, though its efficacy is more on par with that of the pharmacologic options.

Psychotherapy

Psychotherapy has often taken a back seat in treatment-resistant depression, but that is starting to change. Psychotherapy has biological effects, and trials in the past decade have proved its merits in severe and treatment-resistant cases. Not all therapies work here, and the ones that do require a more active role on the part of therapist and patient, such as in behavioral activation and intensive short-term psychodynamic psychotherapy (Town JM et al, *J Affect Disord* 2017;214:15–25).

Limitations of Current Research

That's a lot of progress, but there is still a lot that is missing. There's a lack of controlled trials in patients with high levels of treatment resistance. In fact, only 30% of augmentation trials enrolled patients with true treatment resistance (≥ 2 failed antidepressants). The rest looked at what happens after a single failed antidepressant.

Clinical trials also miss the long-term view. Difficult-to-treat depression is a chronic illness, but most trials only tell us what happens over a few months. Long-term trials are expensive, and keeping patients on a placebo for years is ethically questionable. I'll fill those gaps as best I can in this book, often with data from large observational studies. These give a longer view, but their lack of randomization makes them open to biases that are only partly controlled with statistical techniques.

In the next chapters, we'll look at two switch strategies that may bring some patients to recovery: MAOIs and tricyclics.

Key Takeaways

- Two-thirds of patients do not recover on their first antidepressant.
- Augmentation is generally more effective than switching antidepressants, while TMS and ECT are more effective than augmentation.
- First-line augmentation strategies include lithium, second-generation antipsychotics, pramipexole, and the ketamines.
- Dose increases typically don't improve response beyond the medium dose range for most antidepressants (exceptions: tricyclics, MAOIs, desvenlafaxine).

CHAPTER 16

Monoamine Oxidase Inhibitors (MAOIs)

THE MONOAMINE OXIDASE INHIBITORS (MAOIS) were originally developed in the early 1950s to treat tuberculosis. Some patients experienced unexpected relief from depression on them, leading to their discovery as antidepressants. It was a momentous discovery, but it was not viewed that way at the time. "This is not the first drug which has proved successful in the treatment of depressive conditions," wrote Heinz Lehmann as he welcomed the MAOI iproniazid to a long list of pharmacologic options that were used for depression at the time: amphetamines, opium, barbiturates, vitamin B3, testosterone, and hematoporphyrin (Deverteuil RL and Lehmann HE, *Can Med Assoc J* 1958;78(2):131–133).

The tricyclics joined that list soon after, and by the early 1960s psychiatrists recognized these newer agents as more reliable than their predecessors. They knew that atypical depressions responded better to MAOIs and melancholic to tricyclics, and they also knew about their risks. Both drugs could trigger mania, both had withdrawal phenomena, both could be fatal in overdose, but MAOIs caused two additional risks that held them back.

When used on their own, MAOIs are reasonably well tolerated, but they interact dangerously with tyramine-rich foods like aged cheese (hypertensive crisis) and serotonergic medications (serotonin syndrome). Without these risks, the MAOIs may have eclipsed the tricyclics, as they are otherwise the better tolerated of the two and more effective for the atypical features that are common in outpatient practice.

Evidence in Depression

MAOIs are effective in depression, but whether they work in treatment-resistant cases is less clear. Here the clinical lore is high, but the evidence is low, hovering at the level of open-label case series. Most impressive is a series of patients who responded to high-dose tranylcypromine (Parnate) (90–170 mg/day) after an average of eight antidepressant therapies failed. High-dose phenelzine (Nardil) is not as well studied in treatment-resistant depression but has been tried in doses up to 120 mg/day (Amsterdam JD and Shults J, *J Affect Disord* 2005;89(1–3):183–188; Amsterdam JD and Berwish NJ, *Pharmacopsychiatry* 1989;22(1):21–25). Of the seven patients in that trial, four had complete remission, one had a partial response, and the other two did not respond. Other trials found encouraging response rates to MAOIs (around 50%) in patients who did not respond to tricyclic antidepressants (Ricken R et al, *Eur Neuropsychopharmacol* 2017;27(8):714–731).

In 2006, the FDA approved a transdermal version of an MAOI that was originally developed for Parkinson's disease, selegiline (Emsam). Though it is better tolerated than the older MAOIs, no trials have tested selegiline in treatment-resistant depression except in its oral form, and that study used higher doses than what is available with the transdermal patch (20–60 mg/day, which equates to 6–18 mg/day of transdermal patch).

When to Consider an MAOI

Patients who respond well to MAOIs include women and those with atypical symptoms (over-eating, oversleeping, leaden paralysis, and rejection sensitivity). Consider a high-dose MAOI for patients with high levels of treatment resistance.

MAOIs also treat panic disorder (particularly tranylcypromine) and social anxiety disorder (particularly phenelzine), and may be more effective in those anxiety disorders than the selective serotonin reuptake inhibitors (SSRIs).

Mechanism of Action

MAOIs increase the three monoamines involved in depression: norepinephrine, serotonin, and dopamine. They do this by inhibiting the MAO-A and MAO-B enzymes that metabolize these neurotransmitters. The older MAOIs— tranylcypromine, phenelzine, and isocarboxazid (Marplan)—cause irreversible inhibition at these enzymes. The newer MAOI, transdermal selegiline, is also irreversible, but it has a slightly different pharmacology. At low doses (6 mg/day), transdermal selegiline is selective for MAO-B, which primarily increases dopamine. At higher doses (≥ 9 mg/day), it inhibits both enzymes.

How to Use MAOIs

Selecting an MAOI

There are four MAOIs to choose from, and each has its role. For treatment-resistant depression, I recommend the older MAOIs. Among them, tranylcypromine is often preferred; it is less sedating and there is better evidence to guide us, particularly in the high doses. Phenelzine is better for anxious depression and those who find tranylcypromine too activating. Isocarboxazid is a third-line strategy because it is the least studied of the MAOIs and has a reputation as the least effective (earning it the nickname phenelzine-lite).

Selegiline is a good choice for patients who have trouble tolerating traditional MAOIs. The lack of food interactions in the low dose is also a plus, but I find that most patients who need an MAOI end up needing a higher dose (observational reports suggest MAOIs have a linear dose-response, but studies on this are lacking).

Dosing and Titration

Start at the lowest dose and titrate slowly, raising every three to seven days as described in the table below. Many patients require doses at the higher end of the range to achieve full remission, particularly in treatment-resistant cases.

Side Effects

Tolerability

MAOIs are about as well tolerated as other antidepressants, with fewer gastrointestinal problems than SSRIs but a greater risk of hypotension. Rarely, they cause agitation, insomnia, weight gain, and sexual dysfunction. Selegiline patch is even better tolerated, with minimal weight gain, fatigue, and sexual dysfunction. The patch can cause skin irritation, which can be managed by rotating the application site or using hydrocortisone cream (allow it to dry before applying the patch).

TABLE 16-1. MAOI Antidepressants

Name	Dosing	Common Adverse Effects
Tranylcypromine (Parnate)	Start 10 mg once or twice daily, titrate every 5–7 days to 30–60 mg/day, divided BID	Sexual dysfunction, orthostatic hypotension, insomnia, agitation, peripheral edema
Isocarboxazid (Marplan)	Start 10 mg QD, titrate every 5–7 days to 30–60 mg/day, divided BID	Sexual dysfunction, orthostatic hypotension, weight gain, insomnia, sedation, peripheral edema, vitamin B6 deficiency
Phenelzine (Nardil)	Start 15 mg once or twice daily, titrate every 5–7 days to 45 mg-90 mg/day	Sexual dysfunction, weight gain, orthostatic hypotension, insomnia, sedation, peripheral edema, vitamin B6 deficiency
Selegiline transdermal (Emsam)	Start 6 mg QD transdermal, raise if incomplete response to 9–12 mg QD	Insomnia, topical reactions at patch site, false-positive on drug screen for methamphetamine

Among the older MAOIs, tranylcypromine causes less weight gain and sedation, possibly because of its more dopaminergic properties, with a greater affinity for MAO-B than MAO-A. Tranylcypromine can cause transient increases in blood pressure a few hours after the dose, presenting as palpitations and headaches.

Risks

MAOIs can induce mania, but the studies on this are sparse so the risk is difficult to estimate. Their medical risks include two drug interactions (hypertensive crisis and serotonin syndrome), hypotension, edema, vitamin B6 deficiency, and—rarely—seizures and hepatotoxicity.

Hypertensive Crisis

Hypertensive crisis is a sudden spike in blood pressure caused by the combination of MAOIs with tyramine-rich foods like aged cheese. While everyday hypertension is a "silent killer," slowly damaging the body without causing any symptoms, the sudden hypertensive crisis brought on by this interaction is not silent. Vessels can burst, causing end-organ damage and stroke. Signs of hypertensive crisis include severe headache, confusion, blurry vision, chest pain, and seizures.

Tyramine causes this reaction by raising norepinephrine. Normally, the MAO-A enzyme breaks that norepinephrine down, but with an irreversible inhibitor on board it can't. In theory, noradrenergic medications like amphetamines could also cause this, but there are no cases of hypertensive crisis when they are combined with MAOIs. However, amphetamines may cause serotonin syndrome, as we'll see next.

In the past, hypertensive crises were difficult to predict because we lacked accurate technology to measure tyramine in foods. The dangerous uncertainty led to over-expansive restrictions. Few patients were willing to sign up for a life devoid of pizza, beer, and red wine. That started to change in the 1990s, culminating in a study that found safe levels of tyramine in pizzas from major chains, including four slices of a double-cheese, double-pepperoni Domino's pizza (Shulman KI and Walker SE, *J Clin Psychiatry* 1999;60(3):191–193). These food interactions are more manageable with the Modernized MAOI Diet in the table below.

Another way to reduce this risk is with the selegiline patch. In theory, its transdermal delivery minimizes tyramine inhibition in the gut, but this has only been tested at the low dose (6 mg/day). That dose is free of food restrictions, making it easy for patients to start, but patients will still need to adopt the MAOI diet if the dose is raised to 9–12 mg, where the more dangerous inhibition of MAO-A kicks in.

Serotonin Syndrome

Many psychiatric medications carry a risk of serotonin syndrome, but it's not something we walk on pins and needles about. Not so with MAOIs. Their irreversible inhibition of the enzyme that breaks down serotonin means this interaction is more common and more serious with MAOIs. Low-dose selegiline may be free of food interactions, but it is not free of serotonin syndrome.

The table below lists drugs to avoid with MAOIs. Besides the usual round of antidepressants, there are opioids, psychedelics, ketamines, dextromethorphan, lithium, ADHD medications (stimulants and viloxazine [Qelbree]), and two antipsychotics (ziprasidone [Geodon] and lumateperone [Caplyta]) to worry about. Just as important is to know which drugs do not pose a risk. That includes other antipsychotics, nonserotonergic opioids (morphine, codeine, oxycodone, and buprenorphine [Suboxone]), and lavender oil (Silexan).

Carbamazepine (Equetro) and the anti-migraine triptans appear on older lists for theoretical reasons, but these have since established a safe track record. Also missing from the list are the alpha-antagonists like clonidine (Kapvay) and guanfacine (Intuniv). These interact with MAOIs by lowering blood pressure, but they do not cause serotonin syndrome. More controversial is whether MAOIs can be combined with noradrenergic antidepressants (see side bar on page 128).

To prevent serotonin syndrome, a washout period is required between stopping an antidepressant and replacing it with an MAOI (even when switching from one MAOI to another). Generally, the old drug should be tapered down over one to two weeks (fluoxetine and vortioxetine can be stopped abruptly because of their long half-lives). Once the old drug is completely stopped, you should allow for a washout period equivalent to five half-lives of the old drug before starting the MAOI. Benzodiazepines can be used to relieve any distress that arises during that drug-free period.

TABLE 16-2. Medications that Can Cause Serotonin Syndrome with MAOIs

Serotonergic Psychotropics	Serotonergic antidepressants (SSRIs, SNRIs, other MAOIs, vortioxetine, vilazodone, clomipramine, imipramine, and possibly trazodone and nefazodone); viloxazine, dextromethorphan (Auvelity), ketamine, esketamine, lumateperone, ziprasidone, and possibly buspirone, gepirone, and lithium
Stimulants	Amphetamine, methylphenidate, phentermine, and local anesthetics that contain sympathomimetics
Serotonergic Opioids	Fentanyl, methadone, meperidine, oxycodone, propoxyphene, tramadol
Other	Fenfluramine, linezolid, methylene blue, moclobemide
Over-the-Counter	L-tryptophan, SAMe, St. John's wort, and decongestants containing phenylephrine, pseudoephedrine, dextromethorphan, brompheniramine, or chlorpheniramine
Drugs of Misuse	Cocaine, amphetamines, LSD, MDMA, Ecstasy, bath salts

*These meds can also cause serotonin syndrome with SSRIs and SNRIs, but the risk with these combinations is much lower

Hypotension

MAOIs lower blood pressure through complex effects on the noradrenergic system. The result is syncope and falls, particularly when standing up abruptly. The risk is greater in patients who are older, have poor vasomotor tone, or are taking antihypertensives. Remedies include dose reduction, increasing fluid intake (eight glasses of water a day), and lowering concurrent antihypertensives. When that doesn't work, consider elastic abdominal binders, support stockings, or medications (eg, midodrine 2.5–10 mg TID, fludrocortisone 0.1–0.2 mg QD).

B6 Deficiency and Edema

Phenelzine and isocarboxazid are also associated with vitamin B6 deficiency, which may present as paresthesias, neuropathy, or edema and is treatable with supplementation (vitamin B6 25–50 mg QD).

Controversy: Augmenting MAOIs with Antidepressants

- Combining a serotonergic antidepressant with an MAOI is potentially lethal, but what about adding a nonserotonergic antidepressant? This strategy is controversial but has become slightly more acceptable as the number of reassuring case series has grown. For example, 81% of patients responded to a combination of tranylcypromine and amitriptyline in a case series of 31 patients who did not respond to electroconvulsive therapy (ECT) (Ferreira-Garcia R et al, *J Clin Psychopharmacol* 2018, 38(5):502–504). In theory, the combination is safer if the MAOI is added to a stable tricyclic rather than the reverse.
- Although that study found no problems with amitriptyline, I recommend augmenting with more noradrenergic tricyclics (nortriptyline and desipramine), and would avoid the ones with the strongest serotonergic effects (clomipramine, amitriptyline, and imipramine). Bupropion, mirtazapine, and trazodone (which has been used successfully to treat MAOI-induced insomnia) are also low risk, as are reboxetine and agomelatine (which are used outside the US).
- Another risky strategy is combining psychostimulants with MAOIs. Serotonin syndrome is possible with this combination, but it is rare, with only a few case reports in the past 50 years. Between the two stimulants, methylphenidate is safer than amphetamine, and the modafinils are safer still. The risk is further reduced by adding stimulants to MAOIs slowly and using low doses.
- Other augmentation agents that can be tried with caution include lithium, triiodothyronine (T3), and pramipexole.
- While these MAOI combination strategies may be appropriate for patients with severe, refractory depression in the hands of careful,

experienced clinicians, they're not for everyday use. At the very least, monitor blood pressure when embarking on them and titrate slowly.

Discontinuation

Sudden withdrawal of MAOIs can disrupt mental health, causing anxiety, agitation, and rare reports of mania, psychosis, and delirium. Taper slowly over at least two weeks. Wait another two weeks after discontinuation before lifting the dietary and drug restrictions or starting a new antidepressant (including another MAOI).

Key Takeaways

- MAOIs are underutilized but effective options for difficult-to-treat depression, particularly in patients with atypical features or comorbid anxiety disorders.
- Higher doses may be necessary for patients with high levels of treatment resistance (up to 170 mg of tranylcypromine or 120 mg of phenelzine).
- MAOIs are reasonably well tolerated but interact dangerously with serotonergic drugs (serotonin syndrome) and tyramine-rich foods (hypertensive crisis).
- The modern MAOI diet is more manageable than commonly believed—many foods previously restricted (like pizza and beer) are safe in reasonable quantities.

TABLE 16-3. A Modernized MAOI Diet

Avoid Completely	• Highly aged cheeses and aged beef (eg, charcuterie boards) • Sourdough bread that is homemade or from a bakery (commercially produced is OK) • Fermented soybean products (found in Asian foods like tempeh, miso, pickled tofu, and bean paste) • Raw meat or fish that has not been refrigerated properly or is past the use by date • Homemade beer or wine

TABLE 16-3. A Modernized MAOI Diet

OK in Small Portions (less than a typical serving size)	• Specialty soy sauce • Dried, aged sausage and salami • Sauerkraut • Beer that is micro-brewed, on tap, or requires refrigeration (limit to 1 standard drink)
OK in Normal Portions (but don't overindulge)	• Cheeses that are not highly aged • Chocolate • Caffeinated beverages • Wine from a commercial producer (no more than 2 glasses) • Beer that is shelf-stable or pasteurized (no more than 2 pints) • Fresh beef or fish • Fava beans • Bananas and avocados that aren't overly ripe • Soy sauce or fish sauce from grocery store brands • Worcestershire sauce • Kimchi • Commercially produced sourdough bread • Fermented yeast products (Marmite and Vegemite)
No Restrictions (barely any tyramine here)	• Milk, yogurt, cream • Nonaged, soft cheese (mozzarella, American, ricotta, cottage cheese, cream cheese) • Dry, cured meats (prosciutto, pepperoni) • Smoked or pickled fish • Fresh chicken, duck, pork, and sausage • Stock cubes, powder, or bullion • Nonfermented soybean products

Source: https://psychotropical.info/wp-content/uploads/2018/02/3_MAOI_Diet_Abbreviated_2016_3.1-1.pdf

CHAPTER 17

Tricyclics

TRICYCLIC ANTIDEPRESSANTS (TCAs) represent one of our oldest options for difficult-to-treat depression. While often overlooked in modern practice, tricyclics offer several advantages for patients who haven't responded to first-line treatments. Their robust noradrenergic effects make them particularly useful for melancholic depression—a subtype that often responds poorly to newer antidepressants. Additionally, their established dose-response relationship at higher therapeutic levels provides clear pathways for optimization when other medications have failed.

Tricyclics were discovered in the 1950s, shortly after the monoamine oxidase inhibitors (MAOIs). Their popularity soon surpassed that of the MAOIs, not because of better efficacy or tolerability, but because they did not interact dangerously with food the way MAOIs did. The tricyclics also had an unfair advantage. Men respond slightly better to this class, and men were overrepresented in the early antidepressant trials. That bias did not end until 1993, when the FDA lifted a rule banning women of childbearing age from drug trials.

Although the tricyclics do not interact dangerously with food, they do pose a serious risk that has caused psychiatrists to reserve them for the most severe depressions. They are fatal in overdose, causing cardiac arrhythmias that ended the lives of many whose depression did not respond to the medications, including singer-songwriter Nick Drake. When the selective serotonin reuptake inhibitors (SSRIs) arrived in the 1980s, their relative lack of toxicity led to their widespread adoption.

Evidence in Depression

When all depressions are lumped together, the tricyclics are equally as effective as other antidepressants (Undurraga J and Baldessarini RJ, *J Psychopharmacol* 2017;31(9):1184–1189). However, tricyclics are potentially more effective for certain conditions.

- **Melancholic depression**—Melancholic depression is characterized by widespread loss of pleasure, ruminative guilt, loss of appetite, early morning awakening, and prominent psychomotor changes. Patients with this subtype of depression respond better to tricyclics than SSRIs (Undurraga J et al, *J Psychopharmacol* 2020;34(12):1355–1341).
- **Depression with chronic pain**—Tricyclics relieve neuropathic pain and fibromyalgia independently of their antidepressant effects. These benefits are comparable to duloxetine's and superior to the SSRIs.
- **Depression with irritable bowel syndrome**—Tricyclics, particularly amitriptyline and imipramine, are more effective than SSRIs in irritable bowel syndrome (Xie C et al, *PLoS One* 2015;10(8):e0127815).

In treatment-resistant depression, tricyclics have mainly been tested as augmentation agents, particularly with SSRIs (Taylor D, *Br J Psychiatry* 1995;167(5):575–580). In theory, the tricyclics provide additional noradrenergic effects that enhance the antidepressant response, but this combination can also be dangerous. Many modern antidepressants raise tricyclic levels,* and high levels can cause fatal arrhythmias (Palaniyappan L et al, *Adv Psychiatr Treat* 2009;15(2):90–99). Nortriptyline is one of the better studied tricyclics for augmentation, and its well-established serum levels make drug interactions easier to monitor.

When to Consider a Tricyclic

Consider switching to a tricyclic for patients with melancholic features, chronic pain, or irritable bowel syndrome. When patients have a partial response to an SSRI, tricyclic augmentation may be useful, but other augmentation strategies in this book are safer and have better efficacy.

The combination of nortriptyline with lithium is one of the best-proven strategies to prevent depression after electroconvulsive therapy (ECT) (Sackeim HA et al, *JAMA* 2001;285(10):1299–1307). For patients with obsessive-compulsive disorder (OCD), the serotonergic tricyclic clomipramine may be more effective than SSRIs (Sanchez-Meca J et al, *J Anxiety Disord* 2014;28(1):31–44).

Avoid tricyclics in bipolar and borderline personality disorders. These antidepressants have a high risk of causing mania. In borderline personality

---

*Exceptions include citalopram, escitalopram, desvenlafaxine, mirtazapine, trazodone, vilazodone, and vortioxetine.

disorder, they can trigger aggression and disinhibition, and their toxicity in overdose makes them risky in this population (Soloff PH et al, *Psychopharmacol Bull* 1987;23(1):177–181).

Mechanism of Action

Like the SSRI and serotonin-norepinephrine reuptake inhibitor (SNRI) antidepressants, tricyclics block the reuptake of serotonin and norepinephrine. There are two kinds of tricyclics, and they differ in their relative effects on these two neurotransmitters:

- Tertiary amines: serotonin and norepinephrine
- Secondary amines: primarily norepinephrine

Downstream, the tricyclics have neuroprotective effects. They also block histamine receptors, which may account for their anxiolytic and sedative effects. They pose cardiac risks due to their ability to block sodium and calcium channels.

How to Use Tricyclics

Selecting a Tricyclic

It is usually best to start with a secondary amine due to their higher tolerability. For treatment-resistant depression, nortriptyline is often preferred because it is relatively well tolerated and has an established therapeutic range, while desipramine is a close second. However, each tricyclic has unique properties that may make it the right fit for your patient. Clomipramine is best for OCD, doxepin for insomnia, and nortriptyline for those at risk for falls (see table).

Dosing and Titration

As with most antidepressants, slow titration improves tolerability, such as raising the tricyclic dose every 5–7 days. Unlike most other antidepressants, tricyclics are more effective as the dose goes higher within the approved range. That dose-response relationship is even more precise when guided by the serum level.

Serum Level Monitoring

The three tricyclics where serum levels are most useful are imipramine and amitriptyline, which have a linear dose response, and nortriptyline, which has an inverted "U-shaped" response. For nortriptyline, depression is worse when the serum levels are too low (below 50 ng/ml) or too high (above 150 ng/ml) (Amsterdam J et al, *Am J Psychiatry* 1980;137(6):653–662).

TABLE 17-1. Tricyclic Antidepressants: Comprehensive Guide

Medication	Unique Features	Dosing	Therapeutic Serum Range (ng/mL)
Secondary Amines			
Desipramine	Strongest norepinephrine effects, lowest anticholinergic burden, low risk of sedation	Start 25–50 mg QD, target 75–300 mg QD	150–300
Nortriptyline	Lowest risk of orthostasis and weight gain, along with lithium, a good combo to maintain recovery after ECT	Start 25–50 mg QD, target 50–150 mg QD	50–150*
Protriptyline	Least sedating and may be activating	Start 5–10 mg QD, target 15–60 mg QD	70–170
Tertiary Amines			
Amitriptyline	Highest anticholinergic burden, high sedation	Start 25–50 mg QHS, target 150–300 mg QHS	100–250*
Clomipramine	Best for OCD, most serotonergic	Start 25 mg QD, target 150–250 mg QD	70–200 (for clomipramine) 150–300 (for norclomipramine)
Doxepin	Most sedating, approved for insomnia in low doses	Start 25–75 mg QHS, target 150–300 mg QHS (for insomnia, 3–6 mg QHS or use generic liquid 10 mg/ml, 0.3–0.5 ml QHS)	120–250
Imipramine	Best studied for panic disorder and anxious depression	Start 25–50 mg QD, target 150–300 mg QD	150–300*

*Therapeutic ranges with amitriptyline, imipramine, and nortriptyline are well-established

Sources: Goldberg JF and Stahl SM. Practical Pharmacology, 2021; Kaplan & Sadock's Comprehensive Textbook of Psychiatry, 2017.

Side Effects

Tolerability

Like most antidepressants, tricyclics can cause sedation, weight gain, and sexual dysfunction. These risks vary among the different agents and are generally lower with the secondary amines. Outside of that, their main side effects are anticholinergic.

Anticholinergic Effects

Antichlinergic side effects make patients "dry as a bone, red as a beet, blind as a bat, full as a flask, hot as a hare, and mad as a hatter," as the saying goes. They have a slow build, gradually worsening as the patient ages and other anticholinergic medications are added to their regimen. They may present as difficulty reading (from blurred vision), dental problems (from dry mouth), and—in more serious cases—small bowel obstruction from constipation or urinary tract infection from urine retention (see Table). Anticholinergics can also cause confusion and may increase the risk of dementia (Coupland CA et al, *JAMA Intern Med* 2019;179(8):1084–1093).

The problem is rarely due to one medication, and you can check the anticholinergic burden of their entire regimen at acbcalc.com. Among the tricyclics, amitriptyline is the most anticholinergic while nortriptyline and desipramine are the least.

TABLE 17-2. The Human Cost of Anticholinergic Drugs

Anticholinergic Effect	Why it Matters
Dry mouth	Tooth decay, gum inflammation and ulceration, halitosis; poor dental hygiene is a risk factor for depression and dementia
Constipation	Bowel obstruction with potentially fatal paralytic ileus and sepsis
Urinary retention	Urinary tract infections, renal or bladder damage
Dilated pupils	Acute narrow-angle glaucoma, traffic accidents, falls
Impaired papillary accommodation	Inability to read fine print
Increased heart rate	Increased risk of cardiac arrest
Decreased sweating	Hyperthermia
Decreased bronchial secretions	Mucous plugging of small airways, which worsens respiratory illnesses like asthma and bronchitis
Cognitive impairment	Poor memory and concentration; delirium; increased risk of dementia

TABLE 17-3. Managing Tricyclic Side Effects

Side Effect	Management Approach
Constipation	Combination of docusate 100 mg with sennosides 8.6 mg BID (eg, Peri-Colace). Polyethylene glycol (start 17 g daily, titrate based on response). Bethanechol 10–25 mg TID for severe cases.
Dry mouth	Xylitol gum (eg, Spry) TID[1] Biotene products Pilocarpine 4% drop solution (dilute by mixing 1 part med with 3 parts water. Swish solution in mouth for 1 min q8hr prn dry mouth, do not swallow)
Orthostasis	Elastic abdominal binders or support stockings Oral fludrocortisone 0.1–0.2 mg QD or midrodine 0.5 mg TID
Urinary retention	Bethanechol 25 mg TID
Sexual dysfunction	May improve with dose reduction or switching to a different tricyclic; secondary amines are less likely to cause anorgasmia but more likely to cause erectile dysfunction

[1]Xylitol is a natural sugar substitute that protects the teeth from cavity-causing bacteria. Although xylitol gum is often recommended for dry mouth, recent studies have linked xylitol-sweetened beverages to cardiovascular risks.

Risks

Mania, suicide, and cardiotoxicity are the main concerns with tricyclics. Among the antidepressants, the tricyclics carry the highest risk of mania. They are often lethal in overdose, causing arrhythmias and delaying cardiac conduction. Tricyclics should only be dispensed in small quantities to actively suicidal patients.

Drug interactions that raise tricyclic levels can also be lethal (eg, bupropion, duloxetine, and the non-pram SSRIs: fluoxetine, fluvoxamine, paroxetine and sertraline ≥ 150 mg). At therapeutic levels, the main cardiac risk is orthostatic hypotension and associated falls. Tricyclics are associated with rare cases of hepatotoxicity (approximately 4 per 100,000 patient-years).

For patients with difficult-to-treat depression, the risk-benefit calculation may favor tricyclics despite these concerns, especially when other treatments have failed. However, careful monitoring is important, including:

- ECG monitoring for patients over 40 or with cardiac risk factors
- Regular blood pressure checks, especially when titrating
- Attention to suicidal ideation and limiting prescription quantities when necessary

- Monitoring serum levels to avoid toxicity, particularly with augmentation strategies

Key Takeaways

- Tricyclics are underutilized options with a favorable profile in melancholic depression and depression with comorbid pain or irritable bowel syndrome.
- Secondary amines (nortriptyline, desipramine, and protriptyline) are generally better tolerated than tertiary amines (amitriptyline, imipramine, clomipramine).
- Unlike most antidepressants, tricyclics have increased efficacy at higher doses (within their therapeutic range), and serum levels help optimize the dose.
- Their major risks include cardiac effects, anticholinergic side effects, and toxicity in overdose.

Pharmacologic Augmentation

CHAPTER 18

Overview of Pharmacologic Augmentation

WHEN FACED WITH ANTIDEPRESSANT FAILURE, there are four pharmacotherapy options. You can switch, augment, raise the dose, or wait it out. Among these, augmentation is the most effective, but many of the augmentation strategies used in practice are among the least likely to work.

Augmentation strategies that have the most robust evidence of success are lithium, several second-generation antipsychotics (SGAs), pramipexole, and the ketamines. Each has its detractors. The antipsychotics have poor tolerability, serious long-term risks, and little evidence of long-term efficacy. Lithium has strong preventative benefits, but its narrow therapeutic window leaves patients uncomfortably close to a toxic dose. Ketamine and esketamine work quickly, but without a clear exit plan, which means that patients tend to stay on these addictive substances long-term. Pramipexole is well tolerated, but it can rarely drive patients to hedonic excesses like compulsive gambling.

The next tier of options includes thyroid augmentation and several novel agents. These have shown promise in smaller trials or specific populations but the evidence behind them is less robust than it is for first-line options. They include amantadine, celecoxib, d-cycloserine, and minocycline.

Some clinicians prefer augmentation with medications like bupropion, mirtazapine, stimulants, or buspirone. Though popular, each of these augmentation options failed in large, well-designed trials, despite showing promise in trials that were small or lacked a placebo control. Clinicians may be more comfortable with these options because they are familiar, but we shouldn't confuse familiarity with safety. Untreated depression is neither tolerable nor safe.

Non-Pharmacologic Alternatives

If tolerability is the goal, psychotherapy is a better option, as are some complementary and alternative medicine (CAM) treatments like l-methylfolate, probiotics, and light therapy.

Some argue instead for neuromodulation interventions like transcranial magnetic stimulation (TMS) and electroconvulsive therapy (ECT), which are more effective than pharmacologic augmentation. These offer significant benefits for difficult-to-treat cases but come with their own limitations in terms of access, cost, and, in the case of ECT, side effects.

A Roadmap for This Section

This section provides a comprehensive review of augmentation strategies organized by evidence level and clinical utility. We begin with the first-line options that have the strongest evidence: lithium (Chapter 19), antipsychotics (Chapter 20), and pramipexole (Chapter 21). The ketamines also belong in this group, but we'll cover them in a separate section on rapid-acting medications, highlighting their unique role for acute crises.

Next, we explore second-line augmentation strategies: thyroid hormone (Chapter 22) and celecoxib (Chapter 23). Like the first-line options, these have support from randomized controlled trials, but the data supporting them is smaller. We close with third-line strategies: amantadine (Chapter 24), d-cycloserine (Chapter 25), and minocycline (Chapter 26). Here the trials are few and small or, in the case of minocycline, a mix of positive and negative results.

Some of these augmentation strategies can work on their own. I'll point that out along the way as we examine the evidence base, clinical applications, dosing, side effect management, and specific considerations for difficult-to-treat depression.

First-Line Augmentation Strategies

CHAPTER 19

Lithium

LITHIUM SHINES FOR ITS LONG-TERM benefits. This medication has more evidence to prevent depression, suicide, and psychiatric hospitalizations than most other therapies in this book. Lithium is better tolerated than its reputation suggests. Though it carries medical risks, there is growing evidence that it lowers mortality rates in patients with mood disorders (Chen PH et al, *Acta Psychiatr Scand* 2023;147(3):234–247).

Evidence in Depression

In the short term, lithium augments antidepressants with a moderate effect size, similar to that for the antipsychotics (number needed to treat, NNT = 5). Like the antipsychotic trials, most of the lithium trials tested the medication after just one antidepressant failure. Lithium's augmentation trials are smaller than those for the antipsychotics, with support from 12 placebo-controlled trials enrolling 541 patients (Undurraga J et al, *J Psychopharmacol* 2019;33(2):167–176). However, its long-term benefits are sounder.

When to Consider Lithium

Lithium is not for everyone, and in this chapter, I'll attempt to sketch out the ideal candidate. It is a rough sketch. In bipolar disorder, we have a pretty clear picture of the ideal lithium responder: those with classic, euphoric manias or hypomanias that are cleanly separated from depressions and tend to come on after a depression. In major depression, we have only hints, so take these suggestions with a grain of salt.

Recurrent Depression

Lithium's preventative benefits in depression are more robust than its acute effects, making it ideal for patients with recurrent episodes. Those benefits are supported by 21 controlled trials in unipolar depression with an average duration of two years. Unlike other preventative medications, most lithium trials were not biased by commercial support or drug-withdrawal effects. Overall, lithium prevented depression five times better than a placebo (Undurraga J et al 2019).

Lithium is more likely to work in patients with recurrent depression than it is in those with a single episode. In a non-randomized study that tested lithium as monotherapy, 68% of patients with recurrent depression responded to lithium while none of those with nonrecurrent depression responded (Bschor T et al, *J Clin Psychopharmacol* 2013;33(1):38–44).

One study, however, suggests that lithium's durable benefits might be surpassed by quetiapine (Seroquel).

Researchers in the UK randomized 212 patients with treatment-resistant depression to augmentation with either lithium (mean level 0.85 mmol/L) or quetiapine (mean dose 195 mg). The two drugs looked similar over the first two months, but quetiapine was more effective over the long term of this one-year study (Cleare AJ et al, *Lancet Psychiatry* 2025;12(4):276–288). As we'll see in the next chapter, quetiapine has unique benefits among the antipsychotics that might explain these results. It reduces anxiety and improves sleep quality.

Prevention After Electroconvulsive Therapy (ECT) or Hospitalization

Lithium is the most effective pharmacologic strategy to prevent relapse after ECT. In both bipolar and unipolar patients, it cuts this risk in half from 80% to 40% (Lambrichts S et al, *Acta Psychiatr Scand* 2021;143(4):294–306).

Lithium is often avoided in hospitalized patients, probably due to its slow onset and risk of toxicity in overdose. However, lithium may prevent rehospitalization, even in nonbipolar depression. In large observational trials, lithium—but not antidepressants or antipsychotics—was associated with a 50%–75% reduction in rehospitalization rates (Tiihonen J et al, *Lancet Psychiatry* 2017;4(7):547–553; Pompili M et al, *J Affect Disord* 2023;340:245–249).

Bipolar Features

It seems intuitive that lithium works better in patients with soft signs of bipolar disorder, such as mixed features or a family history of bipolar. Some studies hint in this direction, but it is not a rigorous finding. On the other hand, some experts have used this notion to dismiss lithium's benefits in unipolar depression, suggesting that it only worked in older studies, back when the DSM mixed unipolar and bipolar patients together. This myth was debunked by a recent meta-analysis that carefully removed all bipolar subjects without diminishing lithium's efficacy (Undurraga J et al, 2019).

Suicidality

It is alarming that nearly every neuropsychiatric medication carries an FDA warning about suicide. Not so with lithium, which has robust evidence to prevent suicide.

People with mood disorders carry a risk of suicide that's 10–20 times higher than the general population, but when they take lithium, that risk falls to a level that is indistinguishable from the norm. This is true for completed and attempted suicide, in both unipolar and bipolar disorders, and is based on data encompassing half a million person years (Tondo L and Baldessarini RJ, *Curr Psychiatry Rep* 2016;18(9):88; Song J et al, *Am J Psychiatry* 2017;174(8):795–802). Antipsychotic augmentation, in contrast, does not lower the suicide risk (Tsai DH et al, *Br J Psychiatry* 2025;8:1-9).

Those observations are confirmed in randomized controlled trials, where lithium reduced the risk of completed suicide by 60% compared to placebo in a meta-analysis involving 2,400 patients (Smith KA and Cipriani A, *Bipolar Disord* 2017;19(7):575–586).

Lithium's anti-suicide effects are largely independent of its mood benefits. They apply to attempted suicide, completed suicide, and suicidal thoughts. As one patient put it, "I used to think about driving off the road every day . . . until I started lithium."

In practice, clinicians often avoid lithium in suicidal patients out of concern that they will overdose on it. That makes intuitive sense, but it flies in the face of the evidence. Lithium doesn't meaningfully change access to lethal means in a world where most medicine cabinets are stocked with a fatal dose of Tylenol (about 50 pills). When prescribing lithium to suicidal patients, I tell them that lithium is rarely fatal in overdose, although often disabling. If a patient's risk is acute, I engage the family to dispense the medicine one night at a time.

Elderly

Lithium is often avoided in the elderly out of concerns that it will cause medical problems and drug interactions. These are valid concerns, but most studies find that the elderly tolerate lithium well; there are similar renal risks and discontinuation rates in older and younger adults (Osterland SL et al, *Acta Psychiatr Scand* 2023;147(3):267–275; Flapper M et al, *Int J Geriatr Psychiatry* 2021;36(8):1231–1240). In observational studies, older age predicted better acute and long-term responses to lithium augmentation. The target serum levels are lower in the elderly because the blood-brain barrier becomes more porous with age (Buspavanich P et al, *J Affect Disord* 2019;251:136–140; Lambrichts S et al 2021; Christl J et al, Pharmacopsychiatry 2023;56(5):188-196).

Mechanism of Action

Lithium has so many biological effects that it is difficult to figure out which ones are essential. I'll highlight a few that have particular relevance to depression.

Lithium is neuroprotective, protecting brain cells from stress, injury, and illness. Antidepressants also protect the brain, but they do so in limited regions. Lithium's neuroprotective effects cover broader anatomic regions and more biochemical pathways. This may contribute to lithium's preventative benefits in depression.

Lithium's acute benefits may relate to its effects on neuronal signaling, where it reduces neuronal excitability, dampening signaling pathways involving calcium, glutamate, inositol, and phosphatidylinositol. At the neurotransmitter level, lithium modulates serotonergic, dopaminergic, and gamma-aminobutyric acid (GABA) transmission.

Lithium also reduces inflammation, lowering the cascade of inflammatory cytokines by blocking glycogen synthase kinase–3β (GSK–3β). It modulates immune function, causing a benign elevation of neutrophils, which is often seen on routine labs.

One of lithium's unique mechanisms involves the biological clock, which is often misaligned in mood disorders. Lithium normalizes the expression of the circadian rhythm CLOCK genes. Clinically, this means it can stabilize sleep patterns, helping night owls fall asleep earlier, even though it is not directly sedating.

How to Use Lithium

Dosing and Administration

Lithium requires more patient education than the average psychotropic. Many patients are afraid to start it, but that fear does not match up with reality. In one study, patients were more apprehensive of lithium than quetiapine before starting medication, but those who took quetiapine complained of more side effects (McKeown L et al, *J Psychopharmacol* 2022;36(5):557–565).

I start by explaining that lithium is three different medications. There is low-dose lithium, which has preventative effects. Before it was classified as a medicine, low-dose lithium was popular in health spas and in sodas like 7UP and Coca-Cola. Then there is medium-dose lithium (0.6–0.8 mmol/L), which treats depression and is generally well tolerated. Finally, there is high-dose lithium. At this level (0.8–1.2 mmol/L), lithium treats mania, which is where its reputation for problematic side effects comes from.

Lithium is better tolerated when the starting dose is low (150–300 mg) and is raised slowly (every five days toward 600–900 mg). From that point, I'll dose by serum level (0.6–0.8 is ideal for depression in youth and adults; for age 65–79 years: 0.5–0.7 mmol/L; age ≥ 80: 0.4–0.6 mmol/L). Expect a response within two to six weeks of achieving an adequate level. If the patient is fearful of lithium, I may stay at 150 mg for a month or use the liquid form to create an even lower dose.

Dose lithium entirely at night to reduce side effects and renal risks. Most patients prefer the controlled release version, of which there are two available. Lithium CR (Eskalith) peaks at a lower level than lithium ER (Lithobid), which in theory translates to better tolerability. In practice, the choice usually depends on the dose, as lithium CR comes in size 450 mg and ER as 300 and 600 mg.

For some patients, lithium has little acute benefit but profound preventative effects. This presents a challenge, as patients are better at appreciating acute effects than preventative ones. Comparing their mood chart before and after lithium can give them a more meaningful sense of its benefits.

Lithium is a collaborative effort. It requires thoughtful management on the part of both the patient and provider. Assure your patients that you are committed to making lithium successful and tolerable. When managed well, the extra effort strengthens the therapeutic alliance.

TABLE 19-1. Lithium Levels

Level	Children and Adults	Geriatrics (≥65)	Treats
Low	0.1–0.5	0.1–0.3	Unclear; low levels may augment other mood stabilizers, and might prevent dementia, suicide, and the progression of bipolar disorder
Medium	0.6–0.8	0.4–0.7	Depression (as monotherapy in bipolar depression, or as antidepressant augmentation in unipolar); this is also a good maintenance level to prevent episodes of mania or depression
High	0.8–1.2	0.6–0.9	Active mania
Toxic	>1.2	Varies	Only causes harm

Monitoring

Check labs every 3–6 months on lithium. More frequent checks are needed when patients are starting lithium, psychiatrically unstable, or have risk factors for lithium toxicity, like older age, drug interactions, or declining renal function. Check the lithium level along with basic chemistries, calcium, and thyroid stimulating hormone (TSH).

TABLE 19-2. Lithium Summary

FDA Approval	Bipolar disorder (age 7 and up)
Off-Label Benefits	Antidepressant augmentation, recurrent depression Prevention of suicide, psychiatric hospitalizations, and depression after ECT Medical benefits include prevention of cancer, osteoporosis, viral infections, and possibly dementia and stroke
Dosing	Start 150–300 mg QHS, raise every 5–7 days to target of 600–900 mg QHS then dose by serum level Target level 0.6–0.8 mEq/L Aim 20%–30% lower in older adults
Labs	Check every 3–6 months: TSH, creatinine (eGFR), calcium (more often if older age or drug interactions present) Optional: WBC (benign neutrophil elevation) In vulnerable patients: ECG
Risks	Renal impairment (worse if level > 0.8), hypothyroidism (15%), hyperparathyroidism, nephrogenic diabetes insipidus, psoriasis, cardiac arrhythmias, toxicity

TABLE 19-2. Lithium Summary

Signs of Toxicity	Ataxia, slurred speech, coarse tremor, hyperreflexia, myoclonus, somnolence (typically at levels > 1.2)
Side Effects	Common: Nausea, tremor, thirst, diarrhea. Rare: acne, hair changes, muscle weakness, sexual dysfunction
Half-Life	18–36 hours
Interactions	See Table 19.4

Side Effects

Tolerability

Lithium's tolerability is better than its reputation suggests. Compared to the average antipsychotic, it is five times less likely to cause sedation. Lithium is relatively free of weight-gain. It can cause water retention, but weight gain on the drug is so slight as to be undetectable in the short- and long-term trials (Gomes-da-Costa S et al, *Neurosci Biobehav Rev* 2022;134:104266). Lithium's most common side effect is thirst. Advise patients to drink water and avoid caloric beverages while on it (and avoid diet sodas, which indirectly contribute to weight gain despite their zero-calorie status).

Nausea is common in the beginning. It improves by titrating slowly, switching to a controlled release or liquid form, taking after a large meal, or dividing the dose throughout the day. Rapid intervention is needed. Otherwise, the acute nausea can become conditioned, creating a more intractable scenario of anticipatory nausea. In animal models, that conditioning can be prevented by early intervention with antiemetics. Ondansetron (Zofran) is preferable as it lacks the tardive dyskinesia risk of antiemetics that block dopamine like promethazine (Phenergan) and metoclopramide (Reglan). For patients who do not want to take another medication, ginger capsules are effective for nausea and share a similar mechanism as ondansetron, blocking serotonin 5-HT3 receptors. Patients can further reduce nausea by taking the antiemetic 30–90 minutes before dinner and taking lithium after dinner.

Some patients discover on their own that cannabis treats the nausea, as does its safer cousin, cannabidiol (CBD). Indeed, animal studies of lithium-induced nausea support their experience, but cannabis comes with psychiatric risks (psychosis, apathy, cognitive impairment) and CBD is linked to rare cases of lithium toxicity (Singh RK et al, *Child Neurol Open* 2020;7:2329048X20947896).

Lithium causes a fine tremor that is worse with action, such as drinking or writing. This side effect is dose-dependent and improves with propranolol (Inderal), nimodipine (Nimotop), vitamin B6, or gabapentin (Neurontin) (see table). Reducing caffeine also helps. Patients who need a hot drink can benefit from tremor-proof mugs that reduce the risk of spills (eg, HandSteady).

Lithium's dermatologic side effects include acne, psoriasis, and thinning or coarsening of the hair. Rarely, lithium causes taste changes or muscle weakness.

Most of lithium's tolerability problems are manageable, often with treatments that have psychiatric benefits of their own (see table).

TABLE 19-3. Treatments for Lithium Side Effects

Treatment	Benefits	Potential Psych Benefits	Dose
Propranolol	Tremor	Anxiety	80–240 mg/day
Vitamin B6	Tremor	Depression	500–1,000 mg/day*
Nimodipine	Tremor	Vascular depression, bipolar disorder	240–480 mg/day, divided TID
Gabapentin	Tremor	Social anxiety, alcohol/ cannabis use disorders	600–1,200 mg/day divided BID-TID
Ondansetron	Nausea	OCD, binge drinking, bulimia	4 mg q 12 hr prn
Ginger	Nausea	None	1,000–2,000 mg q12 hr prn
Promethazine	Nausea	None	25 mg q12 hour prn
Metoclopramide	Nausea	None	5–20 mg q4–6 hour prn
Amiloride	Nephrogenic diabetes insipidus	None	5 md/day
Aspirin	Sexual dysfunction in men	None	240 mg/day
Minocycline	Acne	Depression	100–200 md/day
Probiotics	Acne	Depression, anxiety	1 capsule/day
Omega-3	Acne, psoriasis	Depression	2,000–3,600 mg/ day of EPA + DHA
Inositol	Psoriasis	Depression, bulimia	12,000–18,000 mg/ day
N-acetylcysteine (NAC)	Renal protection	Depression	1,200–2,000 mg/ day divided BID

*Vitamin B6 carries a risk of neuropathy that may be avoided by using active form, pyridoxal 5'-phosphate, 200–500 mg, instead of pyridoxine 500–1,000 mg

Risks

Lithium has several medical risks, but it is associated with lower mortality rates than other mood stabilizers. Lithium also has medical benefits. It lowers the risk of cancer, osteoporosis, viral infections, and possibly dementia and stroke.

Renal Effects

Lithium's main risk is renal insufficiency. Recent studies have brought reassuring news about this risk. When viewed through the lens of observational studies involving millions of patients followed for 10 or more years, lithium is no more nephrotoxic than valproate (Depakote) or lamotrigine (Lamictal) (Kessing LV et al, *Eur Neuropsychopharmacol* 2024;84:48–56; Bosi A et al, *JAMA Netw Open* 2023;6(7):e2322056). When that risk is parsed out further, lithium does not seem to harm the kidneys as long as the level is kept below 0.8 mmol/L, but renal function declines with each spike above that level (Clos S et al, *Lancet Psychiatry* 2015;2(12):1075–1083).

Lithium's renal risks are further reduced by dosing it entirely at night and staying on top of the labs (Gitlin M, *Int J Bipolar Disord* 2016;4(1):27). Consult a nephrologist if the creatinine clearance falls below 60 mL/min (this corresponds roughly to a creatinine of 1.5 mg/dL, but the creatinine clearance is more sensitive as it adjusts for weight, age, and gender). A rapid drop in creatinine clearance of more than 25% below baseline also warrants a consult.

A promising strategy for renal protection on lithium is n-acetylcysteine (NAC). This antioxidant prevents glomerular decline in chronic kidney disease. In human studies, NAC prevented nephrotoxicity from vancomycin, amphotericin, and contrast dye, but support for its protective effects against lithium is limited to animal studies (Hernández-Cruz EY et al, *Kidney Int Rep* 2024;9(10):2883–2903). Nonetheless, NAC has psychiatric benefits that may justify its use with lithium. It reduces depression, addictions, and obsessive-compulsive symptoms (particularly trichotillomania) at doses similar to those used in renal studies (1,200–2,000 mg/day).

Lithium can also reduce the kidney's ability to concentrate urine by causing nephrogenic diabetes insipidus (NDI). Patients complain of copious urine production that gets in the way of everyday activities. If left untreated, NDI raises the risk of renal insufficiency. NDI is best managed in consultation with the primary care clinician. The diagnosis requires urine and serum samples and may require fluid restriction and 24-hour urine collection. Urine is dilute, with low osmolality and sodium, while serum sodium levels may be elevated. The first-line treatment is amiloride 5 mg daily, a potassium-sparing diuretic and antihypertensive. Amiloride does not interact with lithium, but it can raise potassium levels, which should be checked along with routine lithium labs.

Thyroid and Parathyroid Effects

The most common medical problem on lithium is hypothyroidism, with a lifetime risk of 10%–20%. Even subclinical hypothyroidism may be worth treating with a low dose of levothyroxine (eg, 25 mcg). In a controlled trial of bipolar depression, lithium worked best when the TSH was kept closer to the middle of the normal range (2.4 mIU/L) (Frye MA et al, *Acta Psychiatr Scand* 2009;120(1):10–13).

Hyperparathyroidism is a separate risk that occurs in about 1 in 25 patients on lithium. Patients may complain of fatigue or brain fog, but most cases are asymptomatic. To screen for hyperparathyroidism, check calcium along with routine lithium labs. If elevated, check the parathyroid hormone (PTH). If that is elevated, refer to endocrinology for further management, which may involve surgery or treatment with cinacalcet (Sensipar). Untreated hyperparathyroidism can progress to osteoporosis and renal stones.

Toxicity and Drug Interactions

Lithium's therapeutic levels sit uncomfortably close to its toxic range. Toxicity causes imbalance, severe tremor, vomiting, slurred speech, and confusion, but its long-term implications are the real concern. Renal function declines with each toxic level and lithium toxicity can damage the cerebellum, causing permanent disturbance of gait. Serum levels inform the assessment, but they are not diagnostic. A young patient may have no signs of toxicity at a level of 1.2 mmol/L, while a level of 0.8 mmol/L may be toxic in an older patient.

Toxicity is not a contraindication to future trials of lithium, but it is a call to clarify the cause and prevent future occurrences. Factors that alter lithium levels are listed in the tables, and most important among these are drug interactions.

Nonsteroidal anti-inflammatory drugs (NSAIDs), antihypertensives, and some antibiotics raise lithium levels. Among the NSAIDs, aspirin is safe, and sulindac (Clinoril) is low risk. Drug interaction checkers flag many other medications that may increase lithium's side effects but do not actually raise its levels, so check first if the interaction is a pharmacokinetic, meaning it alters levels.

Many antipsychotics carry a warning about neurotoxicity with lithium. This unfortunate choice of words is not meant to suggest neuronal damage. Rather, it refers to a rare syndrome of severe tremor, extrapyramidal symptoms, and mental status change when lithium and antipsychotics are taken together.

TABLE 19-4. Factors that Alter Lithium Levels

Raises lithium	Lowers lithium
Drug Interactions: Antihypertensives, diuretics, NSAIDs, and more (see Table 19-5)	Drug Interactions: Acetazolamide, xanthines (aminophylline, theophylline), mannitol
Dehydration	Caffeine
Aging	Active mania*
Renal slowing	Pregnancy
Low sodium diet	Going off a low sodium diet

*Lithium levels fall during active mania, possibly due to increased urination in the manic state.

TABLE 19-5. Medications that Raise Lithium

Class	Examples
Thiazides and loop diuretics	The "—ides": bumetanide, chlorothiazide, furosemide, hydrochlorothiazide (potassium-sparing diuretics like amiloride and spironolactone are OK)
ACE-inhibitors	The "—prils": benazepril, captopril, enalapril, lisinopril
Angiotensin II antagonists	The "—sartans": azilsartan, losartan, valsartan
NSAIDs and COX-2 inhibitors	Over-the-counter: Ibuprofen, naproxen (aspirin is OK) Prescription: Celecoxib, diclofenac, indomethacin, meloxicam (sulindac is usually OK)
Antibiotics	Metronidazole, tetracycline

Lithium Orotate

- Prescription lithium is available as two salts, carbonate and citrate, but a third preparation is gaining popularity as an over-the-counter supplement: lithium orotate. Lithium orotate proponents claim it delivers more lithium to the brain with less renal toxicity, but the evidence is limited to animal studies with conflicting results.

- Despite claims of greater safety, lithium orotate has caused symptoms of toxicity at surprisingly low serum levels in case reports. Most lithium orotate products are sold as a daily dose of 120 mg (equal to 24 mg lithium carbonate or about 5 mg of elemental lithium), similar to levels found in lithium-rich drinking water. While unlikely to be harmful at these low doses, lithium orotate hasn't been sufficiently studied for therapeutic use in depression.
- I wouldn't recommend orotate for treatment of difficult-to-treat depression. For patients interested in lithium but fearful of prescription forms, starting with low-dose prescription lithium (150 mg) is a safer option.

Key Takeaways

- Lithium is ideal for recurrent depression, with better preventative effects than other augmentation strategies.
- Lithium prevents psychiatric hospitalizations, suicide, and depression after ECT.
- Despite its medical risks, lithium has medical benefits and is associated with lower mortality rates.
- Lithium has low rates of sedation and weight gain. Most side effects are manageable, and new research suggests its renal risks are low at the levels used for depression (0.6–0.8 mmol/L).
- Drug interactions and toxicity are serious risks with lithium.

CHAPTER 20

Antipsychotics

ANTIPSYCHOTIC AUGMENTATION HAS RISEN in popularity in recent decades, starting with the approval of aripiprazole (Abilify) for this use in 2007. Seven years later, aripiprazole was the most profitable drug in all of medicine. Approvals of olanzapine-fluoxetine (Symbyax), quetiapine (Seroquel), brexpiprazole (Rexulti), cariprazine (Vraylar), and lumateperone (Caplyta). These augmentation strategies work quickly, but their long-term benefits are less clear.

Antipsychotic augmentation was first deployed in the 1960s to speed up antidepressants and target symptoms of anxiety and agitation. The first combination pill was released in 1961 as Parstelin, pairing trifluoperazine with the monoamine oxidase inhibitor (MAOI) tranylcypromine, followed later by an amitriptyline-perphenazine combination. At the time, they were seen as symptomatic treatments suited to short-term use. Antipsychotics did not prevent depression, and there was a concern that long-term blockade of dopamine could cause depression. To reduce their risks, patients with mood disorders came on and off antipsychotics, but that intermittent use actually raised the rates of tardive dyskinesia.

By the 1980s, the strategy was in decline, a trend predicted by one of the first papers on the antidepressant-antipsychotic combination. "Reports on drug research usually follow a pattern: after extensive and widespread usage of the drug, toxic or allergic side effects are noted, no matter how innocent the chemical may initially have been claimed to be." (Straker M, *Can Med Assoc J* 1960, 83(25):1306–1310).

Today's antipsychotics are not markedly different from the older generation. They work quickly, but most lack preventative effects. Like their forbearers, they are particularly effective for the anxious, agitated symptoms of mixed depression. They cause tardive dyskinesia at a lower rate than the first-generation versions, but there is a higher risk of metabolic problems when using many second-generation antipsychotics (SGAs).

Evidence in Depression

For patients with difficult-to-treat depression, antipsychotics offer several distinct advantages:

- **Rapid onset of action**—Most show benefits within one to two weeks, compared to the four to six weeks typically needed for antidepressants
- **Effectiveness after multiple treatment failures**—There is evidence that some (particularly aripiprazole) are effective even after two to four antidepressant failures
- **Efficacy for complex symptom profiles**—Particularly valuable for depression with mixed features, anxiety, agitation, or psychotic symptoms
- **Different mechanism of action**—May work through pathways not targeted by traditional antidepressants

These benefits must be balanced against significant tolerability concerns and long-term risks that often limit their utility in chronic depression. Careful patient selection and strategic implementation are essential to maximize benefit while minimizing harm.

When to Consider an Antipsychotic

Consider an antipsychotic when your patient needs rapid improvement or has mixed features. They are also likely to work for:

- **Depression with prominent anxiety**—Especially quetiapine, which treats generalized anxiety disorder (GAD)
- **Depression with insomnia**—Although many antipsychotics are sedating, only quetiapine and olanzapine improve sleep quality
- **Psychotic depression**—These cases require higher doses, in the ranges used for schizophrenia

When antipsychotics work, they work fast, within 3 to 14 days. On the other hand, their tolerability problems cause many to reject them. While discontinuation rates are relatively low in the rarified world of clinical trials (around 1 in 10 stopped them in the depression trials), real-world rates of non-adherence are much higher (45%–85%) (Gao K et al, *J Clin Psychiatry* 2011;72(8):1063–1071; Khandker R et al, *J Med Econ* 2023;26(1):878–885). Nausea, falls, akathisia, and sedation are common causes of early discontinuation.

Antipsychotic augmentation works equally well in older and younger adults, but the risk profile differs by age. Older patients are more vulnerable to tardive dyskinesia, falls, temperature imbalance, and anticholinergic side effects. Metabolic risks, in contrast, are greater for children and adolescents.

Maintenance and Prevention

Antipsychotics work quickly, but do they have long-term benefits? The evidence is mixed. Four large trials tested antipsychotic continuation over six months in 2,308 patients with major depression. The antipsychotic failed in all but one of those, with olanzapine-fluoxetine the sole promising exception. In that study, patients were twice as likely to stay well if they continued olanzapine augmentation, compared to remaining on fluoxetine with a placebo (32% vs 16%). However, most of the relapses occurred in the first month, raising the possibility of a withdrawal effect (Brunner E et al, *Neuropsychopharmacology* 2014;39(11):2549–2559). In the other three trials, antipsychotics failed to prevent depression, including two trials of brexpiprazole and one of risperidone (Rapaport MH et al, *Neuropsychopharmacology* 2006;31(11):2505–2513; Bauer M et al, *Acta Neuropsychiatr* 2019;31(1):27–35; McIntyre RS et al, *Acta Neuropsychiatr* 2024;17:1–12).

Two groups that might benefit from maintenance antipsychotic augmentation are those with psychotic depression and those who didn't respond at all to the antidepressant. In a 9-month trial of patients with psychotic depression, continuing olanzapine (15 mg) lowered the relapse risk from 55% to 20% compared to replacing it with a placebo (all patients stayed on sertraline 150 mg) (Flint AJ et al, *JAMA* 2019;322(7):622–631).

Despite these disappointing results, many patients have a hard time coming off antipsychotic augmentation. Still, it is worth trying, given their long-term risks and lack of clear benefit. To maximize success, wait for at least six months of full recovery and taper slowly over 1–3 months while adding in lifestyle changes to prevent relapse.

How to Use Antipsychotics

Selecting an Antipsychotic

In terms of efficacy, the differences among these agents are minor. Most ring in with a small effect size, around 0.3–0.4, about the same as the effect for selective serotonin reuptake inhibitor (SSRI) antidepressants. If we have to pick a winner, it would be aripiprazole. This antipsychotic has the largest effect size across the most studies, including four trials that involved high levels of treatment resistance (two to four antidepressant failures) (Yan Y et al, *Psychol Med* 2022, 52(12):2224–2231).

For some patients, quetiapine is a better place to start, particularly if they have prominent anxiety or insomnia. While many antipsychotics are sedating, only quetiapine and olanzapine improve sleep quality by promoting deep sleep (ziprasidone and paliperidone have this effect as well, although they lack approval in depression). Quetiapine also has unique efficacy in GAD (50–150 mg/day), with a moderate effect size (0.6) that nearly earned it FDA approval there. Quetiapine has also proved its merit in head-to-head trials with other augmentation agents that lasted up to a year. In those trials, it was slightly more effective than lithium and slightly less effective than esketamine (Cleare AJ et al, *Lancet Psychiatry* 2025, 12(4):276–288; Reif A et al, *N Engl J Med* 2023, 389(14):1298–1309).

While other antipsychotics require an antidepressant to work in unipolar depression, quetiapine works as monotherapy, both in major depressive disorder (MDD) and GAD. The reason may be that quetiapine produces a metabolite with antidepressant-like effects on norepinephrine and serotonin, norquetiapine. Despite sound evidence, the FDA decided not to approve the drug as monotherapy in depression, and not to approve it at all in GAD, because of concerns that it would lead to widespread use of a high-risk medication.

Lumateperone is another good choice for patients who have trouble sleeping. Somnolence is one of the most common side effects with this antipsychotic, which has broad efficacy across the mood spectrum. Although it is not known to work in mania, lumateperone augments antidepressants in unipolar depression and, as monotherapy, it treats bipolar depression and major depression with mixed features.

Olanzapine-fluoxetine might seem like it has an edge, as it is the only antipsychotic with FDA approval in treatment-resistant depression (the others are approved as augmentation). However, its effect size is smaller and its trials did not extend beyond two failed antidepressants.

The antipsychotics differ more in their side effect profiles, and this is often what guides the choice. There is no such thing as a tolerable antipsychotic, but brexpiprazole, cariprazine, and lumateperone are relatively favorable in their class. Aripiprazole is also reasonably well tolerated, although short-term akathisia and long-term weight gain are problems. Start low (2 mg per day or every other day) and raise slowly to reduce the akathisia risk.

Quetiapine is more challenging, with high rates of sedation, orthostasis, and weight gain, but a low risk of akathisia. The XR version reduces orthostasis and eases tolerability during titration. The XR is the only version approved in major depression, but the instant release is also effective there.

Dosing Strategies

Start at the lower end of the dosing range and titrate up based on response and tolerability. The table lists the "optimal doses" where most patients found the best balance of efficacy and tolerability. Going too high can cause side effects that mimic depression, like muscle slowing and fatigue. Here are some principles to guide the dose:

- **Start low**—Particularly important to minimize akathisia, sedation, and hypotension
- **Once-daily dosing**—Most can be dosed once daily, preferably at night to minimize daytime sedation
- **Food requirements**—Lurasidone and ziprasidone require a full meal for proper absorption (at least 350 calories for lurasidone and 500 calories for ziprasidone). Quetiapine XR releases its dose too quickly if taken with food or alcohol, so it is best to take it at least 30–60 minutes after a meal and 12 hours before waking up.
- **Response timeline**—Expect initial effects within one to two weeks; allow four to six weeks at a therapeutic dose before determining efficacy

TABLE 20-1. Second-Generation Antipsychotics for Antidepressant Augmentation*

Medication	Dose Range (mg/day)*	Optimal Dose (mg/day)*	Pros	Cons
FDA Approved				
Aripiprazole	5–15	5	Supported by the most studies, including studies of 2–4 antidepressant failures	Akathisia; long-term weight gain
Brexpiprazole	2–3	2	Effective for comorbid post-traumatic stress disorder (PTSD) with sertraline; favorable tolerability	Small effect size
Cariprazine	1.5–3	1.5	Favorable tolerability	Small effect size
Lumateperone	42	42	Favorable tolerability within its class (FDA approval pending)	Sedation
Olanzapine-fluoxetine combination	6/25-12/50 olanzapine/ fluoxetine	9/45	FDA-approved for treatment-resistant depression (≥ 2 failed trials); effective for depression with mixed features (with fluoxetine)	High rate of metabolic problems; only studied with fluoxetine
Quetiapine	150–300	300	Unique benefits in anxiety and sleep; only antipsychotic with good evidence as monotherapy in major depression; low rates of akathisia	High risk of sedation, orthostasis, and metabolic problems
Off-Label				
Lurasidone	20–60 (with a full meal)	36	Effective for depression with mixed features (as monotherapy)	Only tested in depression with mixed features
Risperidone	0.5–3	1.5	Ranks high in efficacy, along with aripiprazole	High risk of prolactinemia and EPS
Ziprasidone	40–160 (with a full meal, divided BID)	98	Low metabolic risks; treats depression as augmentation and mixed features as monotherapy.	Only two positive trials

*The SGAs omitted from this chart are best avoided in major depression because they are either untested (asenapine, clozapine, iloperidone, and xanomeline-trospium) or failed (pimavanserin).

*Dose ranges are based on clinical trials or FDA approval. Optimal dose derived from dose-response curves (where available) or average levels in clinical trials.

Mechanism of Action

Antipsychotics treat depression in the lower dose range, suggesting that something other than dopamine blockade is responsible for their mood benefits. The list of possible mechanisms is long and varies by agent, including activation of dopamine D3 and serotonin 5-HT1A receptors; inhibition of serotonin 5-HT2A/2C and α2 receptors; modulation of gamma-aminobutyric acid (GABA) and glutamate; as well as decreases in stress hormones (cortisol) and increases in brain-derived neurotrophic factors (BDNF).

Side Effects

Tolerability

Antipsychotics are complex drugs with many side effects. Some of their tolerability problems are related to their medical risks, like dizziness from orthostasis, sexual dysfunction from prolactinemia, and weight gain from metabolic effects. Some cause patients to stop the medication early, like sedation and akathisia, while others build up over time, like tardive dyskinesia and anticholinergic effects. Most of these problems are manageable, and I'll discuss the most important ones below.

Sedation

The first step in managing sedation is to give the entire dose at night. Even short half-life antipsychotics can be given in a single dose without loss of efficacy, although those that require a full meal to ensure absorption present a challenge (eg, lurasidone and ziprasidone). For quetiapine, daytime sedation often is lessened by switching to the instant release version, but the extended-release (XR) is less likely to cause orthostasis.

The next step is to switch to a less sedating agent. This can be like steering between a rock and a hard place, as those that cause less sedation tend to cause more akathisia (eg, lurasidone, risperidone, cariprazine, and aripiprazole), while those with a low akathisia risk tend to be more sedating (eg, clozapine, olanzapine, quetiapine, and ziprasidone). Brexpiprazole and lumateperone are good middle of the road options. They have relatively low risks of sedation and akathisia (iloperidone is also in the middle, but it has not been tested in depression) (Aiken C, *Psychiatric Times* 2021, Vol 38, Issue 4).

There are no known antidotes for sedation on antipsychotics. Modafinil has been tried, without success (Freudenreich O et al, *J Clin Psychiatry* 2009;70(12):1674–1680).

Akathisia

Akathisia is an inner sense of restlessness. Patients shake their legs, complain that they "can't sit still," and experience high anxiety. The feeling is so uncomfortable that it raises the risk of suicide. Akathisia comes on early in the course of treatment and improves by lowering the dose. If that doesn't work, start propranolol or vitamin B6. Both of these brought similar relief in a head-to-head trial, and B6 also helps in treatment of depression, tremor, and hyperprolactinemia (Shams-Alizadeh N et al, *Iran J Pharm Res* 2018;17(Suppl):130–135). Benzodiazepines are also effective, but their psychiatric risks make them a third-line option.

Weight Gain

Weight gain is worse with clozapine, followed by olanzapine, risperidone, and quetiapine. The risk is lower with ziprasidone, lurasidone, and probably lumateperone, but all antipsychotics can cause it (Burschinski A et al, *World Psychiatry* 2023;22(1):116–128). Newer agents often appear to have a lower risk, but this tends to change as more data comes in. Patients who are more likely to gain weight include adolescents, women, Asians, African Americans, Hispanics, patients starting an antipsychotic for the first time, and those who are thinner at baseline.

Still, many patients do not gain a pound, even on high-risk options like olanzapine. Genetic tests are being developed to identify patients at risk. Pending that, a useful strategy is to have the patient weigh themselves naked, in the morning, after going to the bathroom and before eating breakfast, and repeat that procedure one month after starting the antipsychotic. A gain of ≥ 5% of body weight signals trouble ahead and is a good reason to introduce metformin to prevent further weight gain.

A secret about metformin is that it works better when started early, before the weight gain takes on a life of its own. Diet and exercise are also recommended, but they are difficult to adhere to and less effective than metformin (Wu RR et al, *JAMA* 2008;299(2):185–193). Some antipsychotics increase the rewarding properties of comfort foods, and I warn patients to take that seriously. "You need to clear all junk food out of your house, just as an immunocompromised person needs to stay away from sources of infection."

For weight gain on olanzapine, the branded combo Lybalvi hopes to prevent the problem by combining olanzapine with the opioid antagonist samidorphan. Lybalvi's main benefits are preventative. We don't know if it treats obesity after the fact, as it was only tested in patients without obesity (BMI 18–30 kg/m2). Unlike metformin, Lybalvi does not improve insulin sensitivity or triglyceride levels. If Lybalvi is unaffordable, naltrexone has similar properties and preliminary evidence for antipsychotic weight gain (50 mg/day) (Taveira TH et al, *J Psychopharmacol* 2014;28(4):395-400).

For patients who prefer a natural approach, probiotics can help. Antipsychotics cause metabolic problems in part by disrupting the gut microbiome. Probiotics cut weight gain in half and improved insulin resistance in a three-month controlled trial of patients with schizophrenia on olanzapine (when taken with a prebiotic, ie, fiber) (Huang J et al, *Schizophr Bull* 2022;48(4):850–859). Probiotics also have benefits in depression and are described more in Chapter 40: Probiotics.

If those measures do not control the weight gain, consider a glucagon-like peptide–1 (GLP–1) receptor agonist. These work better than metformin once weight gain has taken off, and they should be reserved for that juncture. They are difficult to tolerate (nausea) and carry medical risks (gastroparesis, biliary disease, and pancreatitis). For those reasons, the FDA has approved them only for people with a BMI ≥ 30 kg/m^2 (or who have a BMI ≥ 27 kg/m2 and an obesity related health condition like diabetes, hypertension, or dyslipidemia).

Among the GLP–1 receptor agonists, liraglutide (Saxenda) requires daily injections, while the other two have the advantage of weekly dosing and greater weight loss. When it comes to weight loss, tirzepatide (Zepbound) ranks most effective, followed closely by semaglutide (Wegovy), with liraglutide further behind. GLP–1 agonists reduce addictive behaviors, particularly alcohol and nicotine use, and may improve cognition in mood disorders (Hendershot CS et al, *JAMA Psychiatry* 2025;82(4):395–405; Mansur RB et al, *J Affect Disord* 2017;207:114–120).

Anticholinergic Effects

We discussed the human cost of these drying effects in the chapter on tricyclics (page 135). Among SGAs, the risk is greater with olanzapine and lower with brexpiprazole, lumateperone, and lurasidone.

Risks

There is a long list of medical and neurologic risks associated with antipsychotics. Most of these are long-term risks, so they don't raise alarms

TABLE 20-2. Treatments for Antipsychotic Side Effects

Treatment	Dose	Benefits	Risks
METABOLIC			
Liraglutide (Saxenda) (subcutaneous injection)	Start: 0.6 mg QD (as subcutaneous injection). Target: 3 mg, after increasing by 0.6 mg weekly	Weight loss	Nausea, gastroparesis, biliary disease, pancreatitis
Melatonin	3–5 mg QHS	Weight loss, insomnia	
Metformin	Start: 500 mg QD with food Target: 2,000 mg daily, divided BID or as XR QD	Weight loss, insulin sensitivity, triglycerides, and prolactinemia	B12 deficiency, hypoglycemia, lactic acidosis; nausea improved by taking with food or using XR.
Naltrexone	50 mg QD with food	Weight loss	Nausea; rare liver toxicity; prevents opioids from working.
Probiotics	1 capsule QD (take with a prebiotic, see Chapter 40: Probiotics)	Weight loss, insulin sensitivity	Risks in immunocompromised
Samidorphan (Lybalvi)	Dosed as combo with olanzapine	Weight loss	Prevents opioids from working
Semaglutide (Wegovy) (subcutaneous injection)	Start: 0.25 mg qWeek. Increase every 4 weeks to target of 2.4 mg qWeek	Weight loss	Nausea, gastroparesis, biliary disease, pancreatitis
Tirzepatide (Zepbound) (subcutaneous injection)	Start: 2.5 mg qWeek. Increase by 2.5 mg to target of 5–15 mg qWeek	Weight loss	Nausea, gastroparesis, biliary disease, pancreatitis
Topiramate	Start: 25 mg QHS Increase by 25 mg/week to target of 50–200 mg QHS (or divide BID)	Weight loss	Cognitive, renal stones
AKATHISIA			
Benzodiazepines	Varies (eg, lorazepam 1–3 mg/day divided BID or TID)	Akathisia	Sedation, cognition, dependence
Propranolol	Start 20–40 q8hr mg PRN. Switch to ER once daily dose is established (80–240)	Akathisia	Hypotension, bradycardia, Raynaud's phenomenon
Vitamin B6	150–300 mg BID as pyridoxal 5'-phosphate*	Akathisia, tremor, prolactinemia	Neuropathy (dose-dependent)
TARDIVE DYSKINESIA			
Amantadine	Start 100 mg Qam. Increase to 200–300 mg divided BID after 1 week (give in morning and afternoon)	Weight loss, tardive dyskinesia	Rare hallucinations; caution in congestive heart failure

*Vitamin B6 carries a risk of neuropathy that may be avoided by using active form, pyridoxal 5'-phosphate

TABLE 20-2. Treatments for Antipsychotic Side Effects

Treatment	Dose	Benefits	Risks
Deutetrabenazine	Start 6 mg BID, increase by 3 mg BID every week to 24 mg BID (take with food)	Tardive dyskinesia	QTc prolongation (reduced with XR form)
Valbenazine	Start 40 mg QD, raise to 80 mg QD after a week	Tardive dyskinesia	QTc prolongation
Ginkgo (EGb-761, Tebonin form)	80 mg TID	Tardive dyskinesia, extrapyramidal side effects, cognition	
Hyperprolactinemia			
Bromocriptine	2.5 mg QHS	Prolactinemia	Hallucinations
Metformin	Start: 500 mg QD with food Target: 2,000 mg daily, divided BID or as XR QD	Weight loss, insulin sensitivity, triglycerides, and prolactinemia	B12 deficiency, hypoglycemia, lactic acidosis
Vitamin B6	150–300 mg BID as pyridoxal 5'-phosphate	Akathisia, tremor, prolactinemia, extrapyramidal side effects	Neuropathy

*Vitamin B6 carries a risk of neuropathy that may be avoided by using active form, pyridoxal 5'-phosphate

in the depression treatment guidelines that focus on the short-term trials that launched their FDA approvals. The more industry influence among the guideline's authors, the less likely they are to mention these problems (Kavirajan H, *Am J Psychiatry* 2024;181(4):342–345). Among the depression guidelines, only two-thirds discussed the need to monitor for metabolic effects and only one-third mentioned the most significant risk: tardive dyskinesia.

Tardive Dyskinesia

Tardive dyskinesia is not easy to detect. Providers and patients have a tendency to overlook it or explain it away, as in "I always wiggle my fingers when I'm nervous." There's some truth to that denial, as anxiety does intensify the movements (as does sleep deprivation). To detect it, conduct a full Abnormal Involuntary Movement (AIMS) exam twice a year. Look for twitching in the eyes, grimacing of the face, puckering in the lips, writhing of the tongue, and random wiggling in the toes or fingers ("piano player's hands"). The tongue is a naked muscle, hence the best place to observe early signs of tardive dyskinesia. Ask the patient to open wide and let their tongue rest inside their mouth. Then have them write their name in the air with their hand (this distracting gesture can bring out hidden tardive dyskinesia, called reinforcement). Patients can also monitor for tardive dyskinesia on their own with an AI-driven video app (it is currently free as tdscreen.ai but may become commercialized as TDtect Diagnostic by iRx-Reminder). This app outperformed trained, human raters in a randomized controlled trial (Sterns AA et al, *J Clin Psychiatry* 2025;86(3):25m15792).

There is a lower risk of tardive dyskinesia with SGAs than with first-generations (3.9% vs 5.5% per year), but that risk evens out in older adults, where both newer and older agents cause tardive dyskinesia at a rate of approximately 5% per year (Correll CU and Schenk EM, *Curr Opin Psychiatry* 2008;21(2):151–156). That annual rate adds up, so that around 30% will develop tardive dyskinesia after a decade on an SGA.

In schizophrenia, antipsychotic withdrawal is usually not an option, so the best move is to treat the tardive dyskinesia with an antidote or switch to an antipsychotic that does not cause it, like clozapine (which is used in schizophrenia but not depression). In mood disorders, a slow taper over 1–6 months is the first step. The longer a patient has taken the antipsychotic, the slower the taper. Tardive dyskinesia can worsen as the dopamine levels rebalance during antipsychotic withdrawal, but should improve after several months off the drug.

If the tardive dyskinesia requires treatment, start with an FDA-approved agent like deutetrabenazine or valbenazine (the dosing regimen is simpler for valbenazine, and it does not need to be taken with food). After that, amantadine has the best evidence, and this medication has additional benefits for weight loss and depression.

Other Risks

The FDA keeps a thorough tally of antipsychotic risks, and the most relevant of those are in Table 20.3. Antipsychotics that have been on the market longer tend to have longer lists of warnings. Several mention neutropenia, a risk most closely linked to clozapine, followed by quetiapine and olanzapine, both of which are structural analogues of clozapine (Glocker C et al, *J Neural Transm (Vienna)* 2023;130(2):153–163). Many have been implicated in case reports of priapism, although only risperidone and iloperidone have an official priapism warning. Aripiprazole and brexpiprazole

TABLE 20-3. FDA Warnings with Antipsychotics

Risk	Details	Increased or Decreased in . . .
Tardive dyskinesia	Involuntary writhing movements that develop with long-term use, most often in the tongue, face, fingers, and toes	• ↑ Elderly, women, mood disorders • ↑ Patients with akathisia or extrapyramidal side effects
Metabolic syndrome	Obesity, type II diabetes, dyslipidemia	• ↑ Adolescents, women, non-whites • ↑ Clozapine, olanzapine, and quetiapine • ↓ Lumateperone, ziprasidone
Hyperpro-lactinemia	Serum prolactin ≥ 15–20 ng/ml raises risk of breast cancer, osteopenia, and visual loss; presents as breast engorgement, milk production, sexual dysfunction, and menstrual irregularities	• ↑ Risperidone • ↓ Aripiprazole, quetiapine
Orthostatic hypotension	Syncope and falls Drop in systolic by ≥ 20 mmHg or diastolic by ≥ 10 mmHg within 3 minutes of standing	• ↑ Elderly, heart disease, dehydration, pregnancy, alcohol use • ↓ Dividing dose BID or using XR
Arrhythmias	Increased risk of potentially fatal torsade's de point with QTc prolongation	• ↑ Elderly, women • ↑ Heart disease* or electrolyte disturbance (low K or Mg) • ↑ Multiple meds that prolong QTc (check at crediblemeds.org)
Anticholinergic effects	These slowly wear down health and quality of life (see Table 17.2)	• ↑ Elderly • ↑ Multiple anticholinergics (check at acbcalc.com) • ↑ Olanzapine • ↓ Brexpiprazole, lumateperone, and lurasidone
Hyperthermia	Impaired thermoregulatory control	• ↑ Elderly
Neuroleptic Malignant Syndrome	Rare, life-threatening syndrome involving fever, muscle rigidity, tremor, confusion, irregular blood pressure and pulse	
Mortality in dementia	Increased risk of death from stroke in older patients with dementia	

*Cardiac problems that raise the risk of torsade's de point include low left ventricular ejection fraction, left ventricular hypertrophy, ischemia, and slow heart rate carry warnings of compulsive gambling, and case reports implicate fellow dopamine D3 agonist cariprazine in this syndrome as well.

Some warnings have not held up over time, such as cataracts on quetiapine, while others haven't yet caught the attention of the FDA, such as recent reports of liver injury on quetiapine (Ko S et al, *Pharmacoepidemiol Drug Saf* 2023;32(12):1341–1349).

Mortality

When it comes to balancing risks and benefits, mortality is the bottom line. With lithium, we saw evidence that patients live longer on the medication compared to other treatments, but with antipsychotics we see the opposite. Compared to patients with nonpsychotic, treatment-resistant depression who switched antidepressants, those who augmented with an antipsychotic died at a higher rate, usually due to cardiovascular causes. The risk of death is higher as the dose goes up, from 1.2- to 3.5-fold higher. The studies behind this are large and long-term, but are not randomized, leaving open the possibility that factors unrelated to the medication are responsible for the higher risk of death (Tsai DH et al, *Br J Psychiatry* 2025;8:1–9; Pan YJ and Yeh LL, *Aust N Z J Psychiatry* 2023;57(9):1253–1262).

Key Takeaways

- Antipsychotic augmentation works quickly (within one to two weeks) for difficult-to-treat depression, particularly when mixed features are involved.
- Among the effective options, aripiprazole has the strongest evidence, while quetiapine has an edge for anxiety and insomnia.
- Although patients are often kept on antipsychotics long-term, there is little evidence that they prevent depression, and they come with significant long-term risks, including metabolic syndrome and tardive dyskinesia.

CHAPTER 21

Pramipexole

PRAMIPEXOLE (MIRAPEX) IS A DOPAMINE agonist that is highly selective for the D3 receptor. This receptor is involved in the antidepressant actions of aripiprazole, brexpiprazole, and cariprazine. Outside of those, D3 is rarely a target of psychopharmacology. In the pathophysiology of depression, however, D3 is a central player.

D3 receptors are densely packed in the nucleus accumbens, the seat of motivation and reward. When these circuits are overactive, people seek out pleasure at any expense. When they are underactive, people have no motivation to do anything. They are tired, slowed down, and anhedonic. They give up easily (if they start at all), caught between the guilt of avoiding action and the drudgery of taking it on.

This sounds distressing, but "distress" is not the best choice of words for this apathetic state. Patients with anhedonia are less responsive to pleasure and pain. This numbing extends even to their experience of depression. They may tell you that everything is "OK" as their mood is worsening. They may run out of medication and neglect to call for a refill, instead waiting for their next appointment to raise the issue.

Why does dopamine decline? Old age, inflammation, stimulant and cocaine misuse, and chronic stress are common causes. The key phrase there is *chronic stress*. Acute stress tends to raise dopamine, inspiring action and paradoxically improving depression. Under chronic stress, after about six months, dopamine declines and depression sets in.

Evidence in Depression

This dopamine theory motivated clinical trials of pramipexole in treating depression. After its release as a treatment for Parkinson's disease in 1997, researchers noticed improvements in mood as well as movement, something not seen with the dopamine precursor levodopa. In animal models, pramipexole reversed the passive, depressive state that sets in after chronic stress. In two small, controlled studies, it treated bipolar depression better than a placebo with a large effect size (0.77–1.1). The manufacturer tested it as monotherapy in a phase II trial of major depression, where it surpassed placebo and equaled fluoxetine (Aiken CB, *J Clin Psychiatry* 2007;68(8):1230–1236).

Psychiatrists encouraged the manufacturer to press on with phase III trials, but the impending generic release made the cost hard to justify. Pramipexole was shelved as a treatment for depression. It did earn FDA approval in restless leg syndrome (RLS), and improved mood in a large, placebo-controlled trial of patients with RLS and depressive symptoms (Montagna P et al, *Sleep Med* 2011;12(1):34–40).

Putting aside trials of depression in Parkinson's disease or RLS, pramipexole's benefits in depression are supported by seven randomized controlled trials, two of which involved bipolar depression (Tundo A et al, Acta Psychiatr Scand 2019, 140(2):116–125). Pramipexole works as monotherapy and augmentation. Its benefits are large even in treatment-resistant cases (effect size 0.9). Most important, they are durable. Its antidepressant effects were sustained for up to a year in a large, placebo-controlled trial of treatment-resistant depression (Browning M et al, *Lancet Psychiatry* 2025;12(8):579–589).

Pramipexole is one of only a few pharmacologic options with evidence in patients with high degrees of treatment resistance, even if that evidence is only observational. It brought remission after failure of aripiprazole, ECT, and over six antidepressant trials (Tundo A et al, *Biomedicines* 2024;12(9):2064; Fawcett J et al, *Am J Psychiatry* 2016;173(2):107–111). Unlike most therapies, its efficacy does not diminish as the degree of treatment resistance goes up, although patients with high degrees of resistance may require higher doses (eg, 2.5 mg instead of 1.5 mg) (Tundo A et al, *Life (Basel)* 2023;13(4):1043).

When to Consider Pramipexole

Pramipexole is appropriate for patients with treatment-resistant depression and bipolar depression. Candidates who may particularly benefit include those with:

- Anhedonia, amotivation
- High levels of treatment resistance (multiple failed medication or neuromodulation trials)

- Inflammatory markers or conditions
- Comorbid restless leg syndrome
- Soft signs of bipolarity
- Bipolar depression (used with a mood stabilizer)

There is no evidence that pramipexole has the potential for addiction or misuse. Patients who are prone to psychosis or obsessive-compulsive disorder (OCD) may experience worsening of those symptoms on it and are less ideal candidates.

Mechanism of Action

Pramipexole is a selective agonist at the dopamine receptor that regulates hedonic drive (D3) in the nucleus accumbens. There are five dopamine receptors, and pramipexole's affinity for D3 is 7-fold higher than the others, including D2, the receptor involved in psychosis.

Pramipexole belongs to a newer class of dopamine agonists, the non-ergot alkaloids. This class represents a safety advantage over the ergot alkaloids, bromocriptine and cabergoline, which cause valvular heart disease at high rates (20%–25%) (Andrejak M and Tribouilloy C, *Arch Cardiovasc Dis* 2013;106(5):333–339). The newer dopamine agonists appear to be free of that risk, which is caused by activation of 5-HT2B serotonergic receptors on the heart. There are no placebo-controlled trials of the other non-ergot alkaloids, ropinirole and rotigotine patch, in treating depression, but they did improve mood in observational studies and trials of Parkinson's disease. Compared to pramipexole, rotigotine is slightly more selective for D3 and ropinirole is less selective.

Pramipexole should not be confused with other dopaminergic medications like methylphenidate and amphetamine. These block dopamine reuptake, making it available to act on the five dopamine receptors in various regions of the brain. These stimulants improve energy and cognition, while pramipexole causes fatigue and has no effect on cognition.

Unlike pramipexole, the stimulants largely failed in studies of depression. At high doses, the excess dopamine they release causes inflammation in the brain. Pramipexole has the opposite effect. It is neuroprotective within the dopamine system and anti-inflammatory in the brain. Inflammation wears down dopamine transmission, reducing its synthesis, packaging, and release within the central nervous system. This suggests that pramipexole may treat inflammatory depression by restoring dopamine tone and decreasing neuroinflammation, a possibility suggested animal models and a clinical trial (Ventorp F et al, *Psychiatr Res Clin Pract* 2022;4(2):42–47).

How to Use Pramipexole

A slow titration is needed to prevent nausea and orthostasis. When introducing pramipexole:

- Start with 0.125–0.25 mg at bedtime.
- Raise by 0.25 mg every five to seven days.
- Expect slight improvement (10%–20% change) at 0.75 mg. From this dose, titrate faster if the medication is tolerated well, raising to 1.0 mg for a week and then 1.5 mg.
- Most patients require between 1–2 mg, though patients with high levels of treatment resistance may require 2–5 mg/day.
- For nausea, consider ginger (1,000–2,000 mg q12 hr prn), ondansetron (4 mg q12 hr prn), or a proton pump inhibitor (eg, pantoprazole 20–40 mg QD).
- Dose at bedtime to minimize daytime fatigue (rarely, patients find it activating and may prefer morning dosing).

Pramipexole works as both monotherapy and augmentation in unipolar depression, although in bipolar disorder it should be used with a mood stabilizer. Response is typically seen within four to six weeks of reaching a therapeutic dose.

Pramipexole is one of a few neuropsychiatric medications that is excreted unchanged by the kidneys (the others are lithium, paliperidone, gabapentin, pregabalin, acamprosate, and amantadine). It is free of pharmacokinetic drug interactions and does not undergo hepatic metabolism.

Pramipexole is available as an extended-release (XR) version, but this has not been studied in psychiatric populations. The XR was developed for Parkinson's disease, where twice daily dosing is often necessary to maintain its motoric effects, but most psychiatric patients can be treated with a single dose at night.

Side Effects

Tolerability

Pramipexole does not cause weight gain, sexual dysfunction, or cognitive impairment. The most common side effects are nausea and fatigue. At low doses (≤ 1.5 mg) these are not a major obstacle, but they cause more discontinuation in the higher dose range (2.5 mg).

As a dopaminergic medication, various motoric side effects can occur on pramipexole, such as muscle twitching. These are rare, though I have seen them in older adults and patients with a past history of cocaine or methamphetamine use disorder. Pramipexole does not cause tardive dyskinesia.

When used in RLS, pramipexole can shift the restless symptoms from the nighttime to the morning (rebound) or cause them to start earlier in the afternoon (augmentation). These problems are worse with long-term use, affecting around 1 in 10 patients after a year with restless leg syndrome. It is not known if rebound and augmentation occur in mood disorders.

Risks

Pramipexole's main risks are cardiac and neuropsychiatric. On the cardiac side, it can cause orthostasis and falls, a rare problem that is reduced by titrating slowly. Although it does not seem to cause valvular heart problems like the ergot alkaloids, it can cause edema, and there are a few case reports of congestive heart failure on the drug.

Like other D3 agonists (eg, aripiprazole), pramipexole carries a warning that it can cause compulsive gambling. Neurologists call this hedonic homeostatic dysregulation, and the syndrome includes a broad spectrum of pleasure-seeking behaviors. Patients may overspend online, play videogames excessively, or have uncomfortable urges to masturbate at work. Substance use disorders are not part of the syndrome, and neither are manic symptoms. The problem is more common in Parkinson's disease (15%) than mood disorders (1-3%), possibly because depressed patients start out with a lower hedonic drive.

Serious impulsivity is rare and improves with dose reduction. More often, patients report benign compulsivity on pramipexole, such as cleaning the garage or organizing their books alphabetically. Substance use disorders are not part of the syndrome, and neither are manic symptoms. In the bipolar trials, pramipexole did not cause mania (where it was used with a mood stabilizer).

Pramipexole can cause psychotic symptoms by activating the D2 receptor. These are usually mild, such as seeing shadows or hearing sounds. They are rare (0.4%) and more common in the higher dose range (3–5 mg). In the cases I have seen, reality testing was intact and the psychosis resolved with dose reduction or discontinuation. Sudden sleep attacks have been reported on pramipexole in Parkinson's disease, including while driving.

There were early reports of melanoma on pramipexole, but those have since been linked to Parkinson's disease rather than the medication. The FDA removed warnings of melanoma from the prescribing information in 2018.

TABLE 21-1. Pramipexole Summary

FDA Approval	Parkinson's disease Restless leg syndrome
Off-Label Benefits	Major depression, bipolar depression
Dosing	Start 0.25 mg QHS, raise by 0.25 mg every 5–7 days to target of 1–3 mg QHS
Risks	Hypotension, edema Hedonic dysregulation (1-3%), hallucinations (0.4%), sleep attacks
Side Effects	Nausea, fatigue
Half-Life	8–12 hours
Interactions	Pramipexole has few interactions because it is excreted unchanged in the urine; cimetidine raises levels by 50% (inhibits renal secretion)

Key Takeaways

- Pramipexole is a D3-selective dopamine agonist with large and durable benefits in unipolar and bipolar depression.
- It may be particularly beneficial in patients with anhedonia, inflammation, or high levels of treatment resistance.
- Pramipexole is well tolerated with no weight gain, sexual dysfunction, or cognitive impairment, making it a valuable option for patients who have experienced side effects with other treatments.

Second-Line Augmentation Strategies

CHAPTER 22

Thyroid

THYROID AUGMENTATION IS PUZZLING. It is endorsed in practice guidelines for treatment-resistant depression. It has good tolerability, a low cost, and has been used in psychiatry for nearly a hundred years. With qualities like that you'd expect widespread adoption, but several obstacles stand in its way. Most clinicians feel unsure of its medical risks. Thyroid augmentation lacks FDA approval, and its small trials have left some questions about its efficacy.

Evidence in Depression

There are approximately 18 randomized controlled trials of thyroid in depression, involving 1,173 patients. Of those, half were acceleration trials. These looked at whether thyroid could speed up response when started with an antidepressant. The other half were augmentation trials that tested thyroid in patients who did not respond to an antidepressant (Touma KTB et al, *Innov Clin Neurosci* 2017;14(3–4):24–29; Lorentzen R et al, *Acta Psychiatr Scand* 2020;141(4):316–326).

As with most treatments, there is a mix of positive and negative data here, so we turn to meta-analyses to sort it out. Those analyses tell us that thyroid accelerates the effect of antidepressants, but they are less certain about its ability to augment antidepressants in treatment-resistant cases. The problem is that most thyroid trials took place in the 1970s and '80s, when trial design was a work in progress. Many trials don't make it through the meta-analytic filters, and those analyses come up positive or negative depending on what they let in (Zhou X et al, *J Clin Psychiatry* 2015;76(4):e487–e498). This places it a step below the antipsychotics and lithium in terms of the certainty of the evidence, but above many popular strategies like augmenting with antidepressants or stimulants.

T3 or T4?

Thyroid hormone comes in two forms: triiodothyronine (T3, Cytomel) and levothyroxine (T4, Synthroid). Both are synthetic analogues of the natural hormone, and one stands out as the winner for depression: T3. It has more supportive trials than T4 and surpassed T4 in a double-blind comparison (Joffe RT and Singer W, *Psychiatry Res* 1990;32(3):241–251). T3 is the form that is active in the central nervous system. On a practical level, its absorption is not as impeded by food as that of T4.

T4 is preferred for hypothyroidism, including that caused by lithium, as this hormone has more effects throughout the body. About one-third of T4 is converted to T3, but some people (about 1 in 5,000) have difficulty with this conversion.

Mechanism of Action

Low thyroid levels depress mood and impair cognition, but this does not explain why supraphysiologic thyroid treats depression. Theories abound, and we have only hints of the mechanism. One possibility is that a subset of depressed patients has central thyroid resistance. They have difficulty getting thyroid into their brain cells, and supraphysiologic thyroid partly overcomes this resistance.

In support of that theory, depressed patients tolerate high levels of thyroid hormone that cause healthy controls to have anxiety, sweating, and palpitations. The implication is that there is some resistance to the hormone. We can use that insight to personalize the dose, lowering it if physical symptoms of hyperthyroidism show up (Bauer M et al, *J Affect Disord* 2002;68(2–3):285–294).

Thyroid supplementation also enhances serotonin transmission, modulates amygdala networks, and normalizes metabolic energy in the brain.

When to Consider Thyroid

Thyroid augmentation is a second-line strategy for difficult-to-treat depression. As one of the better-tolerated options in this book, it is a good choice for patients who have trouble tolerating medications. Studies suggest thyroid augmentation is more effective in (Altshuler LL et al, *Am J Psychiatry* 2001;158(10):1617–1622):

- Middle-aged women
- Patients with prominent fatigue
- Patients with subclinical hypothyroidism (elevated TSH but normal T3/T4) or elevated antithyroid antibodies

Thyroid can also be started with an antidepressant to speed response, but there are more effective ways to achieve this that we'll discuss in the Rapid-Acting Treatments section.

If the patient is already taking T4 for hypothyroidism, additional T3 can be tried but isn't certain to work. This strategy showed promise in a case series but failed in a randomized trial (Joffe RT et al, *Ann N Y Acad Sci* 2004;1032:287–288).

Thyroid augmentation can be considered in both unipolar and bipolar depression, though the evidence is stronger on the unipolar side.

How to Use Thyroid

When implementing thyroid augmentation:

- Start T3 (triiodothyronine, Cytomel) at 12.5–25 mcg daily.
- Increase every seven days toward a dose of 37.5–50 mcg daily.
- After reaching 50 mcg, allow two to four weeks to assess response.
- Some patients need higher doses (100–150 mcg daily), but these should only be attempted if they tolerate the lower dose and have a partial response on it.
- Patients who are taking carbamazepine may need higher doses (200–300 mcg daily), as this medication lowers thyroid levels.

Monitoring is essential, but signs and symptoms, not just the labs, should guide the dose:

- Monitor pulse during treatment and consider decreasing the dose if it goes above 100 bpm.
- Check a thyroid panel at baseline (TSH, free T3, and free T4).
- Recheck at three months, then six months after reaching the target dose. After that, check labs at least once a year.
- If the TSH falls below 0.5 mIU/L, consider lowering the dose, but this is only necessary if the patient has symptoms of hyperthyroidism.
- Postmenopausal women should have their bone density monitored with their primary care physician at least every two years.

As with most augmentation strategies, we don't know how long to keep people on T3. Consider a gradual taper after six months of recovery, but anecdotally most patients with chronic mood disorders need to stay on the hormone to maintain recovery.

Side Effects

Tolerability

Thyroid augmentation is surprisingly well tolerated. Although it can cause anxiety if the dose goes too high, anxiety generally improves with the thyroid augmentation (Pilhatsch M et al, *Int J Bipolar Disord* 2019;7(1):21).

Potential side effects include:

- Tachycardia
- Hot sensations or sweating
- Headache
- Restlessness
- Anxiety
- Tremor

These side effects usually resolve with dose reduction. Most patients report good tolerability even at doses that would cause symptoms in euthyroid individuals, supporting the theory of central thyroid resistance in depression.

Compared to other augmentation strategies for difficult-to-treat depression, with thyroid there is:

- No weight gain
- No sexual dysfunction
- No sedation
- No risk of tardive dyskinesia or other irreversible side effects
- Low cost

Risks

The main risks of thyroid augmentation involve the bones (osteoporosis) and heart (arrhythmias). These problems did not show up in the depression trials, although one study found a possible loss of bone density in postmenopausal women (but not in premenopausal women) (Kelly T et al, *Prog Neuropsychopharmacol Biol Psychiatry* 2016;71:1–6). To put that risk in perspective, serotonergic antidepressants also cause bone mineral loss in postmenopausal women.

Before starting thyroid augmentation, consider:

- Baseline cardiac status (history of arrhythmias or coronary disease)
- Bone health, particularly in postmenopausal women
- Current thyroid function (both hyper- and hypothyroidism)
- Concurrent medications that might affect thyroid levels

TABLE 22-1. Triiodothyronine Summary

Off-Label Benefits	Depression (unipolar and bipolar)
Dosing	Start T3 (triiodothyronine, Cytomel) 25 mcg QD ×1 week then 50 mcg QD × 1 week; if no response at 50 mcg, raise by 25 mcg/week to max of 150 mcg QD
Duration	Continue for at least six months after recovery (many patients need longer to prevent depression)
Risks	Decreased bone mineral density, cardiac arrhythmias
Side Effects	Anxiety, tachycardia, hot sweats, headache, tremor, restlessness
Half-Life	22–24 hours
Interactions	Bile acid sequestrants and ion exchange resins reduce absorption (take 4 hours apart) Enzyme inducers (phenobarbital, phenytoin, carbamazepine, and rifampin) lower T3 levels
Monitoring	Check thyroid panel at baseline and every 3–12 months Lower dose if resting pulse > 100 bpm or symptoms of hyperthyroidism appear (see Side Effects)

Key Takeaways

- Thyroid augmentation with T3 (triiodothyronine) is a well-established but underutilized strategy for difficult-to-treat depression.
- Best candidates include middle-aged women, patients with fatigue, and those with subclinical hypothyroidism or thyroid antibodies.
- Despite theoretical concerns, clinical evidence shows good tolerability and minimal adverse effects on bone density or cardiac function. It does not cause weight gain, sexual dysfunction, or sedation.

CHAPTER 23

Celecoxib

RELEASED AS CELEBREX IN 1998, celecoxib was the first cyclooxygenase–2 (COX–2) inhibitor to enter the market. It was greeted as a safer alternative to the earlier generation of COX–1 inhibitors like ibuprofen, naproxen, and aspirin, which can interfere with blood clotting and thin the gastric lining. Both COX–1 and COX–2 inhibitors are classified as nonsteroidal anti-inflammatory drugs (NSAIDs). They treat pain by reducing inflammatory prostaglandins, and COX–2 inhibitors are more selective for this effect.

Evidence in Depression

When celecoxib was released, the inflammatory theory of depression was gaining ground. Celecoxib was one of the first treatments to test the theory, improving depression in a placebo-controlled augmentation trial of 40 patients who had not responded to an antidepressant (Müller N et al, *Mol Psychiatry* 2006;11(7):680–684). Two decades and 31 trials later, it remains the best-studied anti-inflammatory for depression. Most of these trials tested celecoxib augmentation in patients who had not responded to an antidepressant (Wang Z et al, *World J Clin Cases* 2022;10(22):7872–7882). In one trial, it worked as monotherapy in postpartum depression, itself an inflammatory state where elevated prostaglandins may contribute to the depressed mood.

Inflammation is also part of the pathophysiology in bipolar disorder, where it contributes to depressive, mixed, and manic episodes. Celecoxib improved each of these phases as an adjunct to mood stabilizers, although the bipolar trials are few in number (6) and small in size (Bavaresco DV et al, *CNS Neurol Disord Drug Targets* 2019;18(1):19–28).

When to Consider Celecoxib

Consider celecoxib in patients with treatment-resistant depression who show evidence of inflammation. Specific inflammatory conditions in which celecoxib treated depression include breast and colorectal cancer, bacterial infections (brucellosis), osteoarthritis, postpartum depression, and those with a high-sensitivity C-reactive protein (hs-CRP) ≥ 3 mg/L. Patients who have disorders for which celecoxib is FDA-indicated, such as arthritis, pain, and dysmenorrhea, are also good candidates.

Mechanism of Action

Celecoxib blocks the prostaglandins that activate inflammation. Prostaglandins raise cytokines like interleukin–1 and interleukin–6. These cytokines make a mess in the brain (neuroinflammation), lowering brain-derived neurotrophic factor (BDNF) and altering the serotonin transporter (SERT). The result is not just depression, but treatment-resistant depression, because the SERT transporter is less responsive to serotonergic antidepressants.

Markers of inflammation tend to decline when patients respond to celecoxib, including interleukins and CRP. What is unknown is whether celecoxib works preferentially in inflammatory depression. One trial suggests it does, finding that celecoxib was uniquely effective in patients with elevated hs-CRP (≥ 3 mg/L) (Sampson E et al, *Brain Behav Immun* 2025;123:43–56).

How to Use Celecoxib

Celecoxib can be used as augmentation or monotherapy. Its half-life is 12 hours, consistent with the twice-a-day dosing strategy used in most trials. When introducing celecoxib for depression:

- Start at 200 mg once daily for one week
- If tolerated, increase to 200 mg twice daily
- Consider adding a proton pump inhibitor (eg, pantoprazole 20–40 mg QD) to reduce the risk of gastric ulcers, especially in high-risk patients

When celecoxib works, its benefits are noticeable within four to eight weeks. Because it is well tolerated, a short-term trial is often worthwhile, but with its long-term medical risks, celecoxib should only be continued if its benefits are clear. Even then, I would attempt to taper it off (eg, over two to four weeks) after six months of recovery. We don't have research on celecoxib's maintenance use, but there is a case report of sustained remission over five years of treatment (Chen CY et al, *Gen Hosp Psychiatry* 2010;32(6);647.e7-9).

Side Effects

Tolerability

Celecoxib is one of the better tolerated options for depression. It does not cause weight gain, sedation, insomnia, sexual dysfunction, or mania. Gastrointestinal distress is possible, but typically less severe than with traditional NSAIDs.

Risks

Celecoxib is associated with several risks that, although rare, raise concerns about its long-term use. Although it is safer for the stomach than other NSAIDs, it does cause gastric ulcers at a rate of 2%–4% per year.

Prostaglandins relax blood vessels, reduce blood clotting, and dilate the airways. Celecoxib interferes with these effects, raising the risks of hypertension and nephrotoxicity (through vascular constriction), heart attacks and stroke (through thrombosis), and asthma (through bronchospasm). People who have had aspirin-induced asthma attacks should not take celecoxib.

Celecoxib is a sulfonamide, which means that patients with sulfa allergies should avoid it. These allergies range from mild hives to life-threatening anaphylaxis and Stevens-Johnson Syndrome. Like most NSAIDs, celecoxib raises lithium levels by decreasing its renal excretion. Rarely, celecoxib can cause anemia or elevate liver enzymes.

Before prescribing celecoxib, consider:

- Cardiovascular risk status (hypertension, history of heart attack or stroke)
- Renal function (assess baseline and monitor)
- Gastrointestinal risk factors (history of ulcers, concurrent steroid use)
- History of sulfa allergies (absolute contraindication)
- Pregnancy status (avoid in pregnancy)
- Concurrent medications, especially lithium

Check labs annually, including complete blood count (CBC), liver function tests, and creatinine.

TABLE 23-1. Celecoxib Summary

FDA Approval	Arthritis (osteoarthritis, rheumatoid arthritis, ankylosing spondylitis), acute pain, primary dysmenorrhea
Off-Label Benefits	Major depression, bipolar depression
Dosing	200 mg BID
Risks	Gastric ulcer, thrombosis (heart attacks, stroke), hypertension, heart failure, hepatotoxicity (1%), renal toxicity, asthma, hepatitis, and anemia (0.2%) Avoid if allergies to sulfa or aspirin (can cause Stevens-Johnson Syndrome, anaphylaxis, and asthma); avoid in pregnancy.
Side Effects	GI distress
Half-Life	11 hours
Interactions	Metabolized through CYP2C9. Inhibits CYP2D6 and raises lithium levels by 20–100%.
Monitoring	CBC, LFT, creatinine annually

Key Takeaways

- Celecoxib is the best-studied anti-inflammatory medication for depression, with 31 clinical trials supporting its use.
- It is particularly useful for patients with inflammatory disorders or high levels of hs-CRP (≥ 3 mg/L).
- Although well tolerated, its long-term medical risks (cardiovascular, renal, GI) make celecoxib more appropriate for time-limited treatment (3–6 months) than for indefinite use.

Third-Line Augmentation Strategies

CHAPTER 24

Amantadine

AMANTADINE (SYMMETREL, GOCOVRI) BEGAN LIFE in 1963 as a treatment for influenza A, but it fell out of use as the virus developed resistance to the drug. Amantadine also has neuropsychiatric effects. Physicians first noticed these in the 1960s, when patients with Parkinson's disease suddenly began moving while taking amantadine for influenza. From there, psychiatrists tested amantadine in tardive dyskinesia, and those trials inspired another chance discovery. Amantadine improved not just involuntary movements but depression as well.

Evidence in Depression

The first clinical trial of amantadine in depression came in 1970. Working in Mexico City, Salvador Vale and colleagues randomized 40 patients with chronic depression and added either amantadine or placebo to their ineffective antidepressant. The drug was twice as effective as the placebo (78% vs 38%). Those promising results were not followed up until the 1990s, when physicians once again noticed improvements in mood while using amantadine to treat viral infections (hepatitis C and Borna virus), observations that were confirmed in small, placebo-controlled trials of patients with viral illness and depression (including bipolar depression).

The next test came in treatment-resistant depression, where amantadine successfully augmented imipramine in two small controlled trials (Raupp-Barcaro IF et al, *Braz J Psychiatry* 2018;40(4):449–458). Beyond that, the only evidence of amantadine's antidepressant effects come from case reports, animal studies, and clinical lore.

As COVID-19 spread, amantadine was once again called on to tackle a viral outbreak. It failed against coronavirus, but succeeded in the psychiatric realm, preventing neuropsychiatric symptoms in acute

COVID and improving energy in long COVID (Rejdak K et al, *Eur J Neurol* 2024;31(1):e16045; Harandi AA et al, *Sci Rep* 2024;14(1):1343).

When to Consider Amantadine

Amantadine is best reserved as a last resort for severe, difficult-to-treat depression. Its limited evidence base and potential risks mean that it should only be considered after more established augmentation strategies have failed.

Outside of depression, amantadine has potential benefits for a variety of conditions that often occur with depression. It can improve energy, cognition, irritability, obsessive-compulsive disorder (OCD), and psychotropic side effects like sexual dysfunction on antidepressants and weight gain and tardive dyskinesia on antipsychotics. Amantadine has preliminary studies in bipolar disorder and a low risk of inducing mania.

Mechanism of Action

As an antiviral agent, amantadine blocks viral replication. Its neuropsychiatric effects are separate. Like ketamine and lamotrigine, it is glutamatergic, antagonizing the N-methyl-d-aspartate (NMDA) receptor. Like the tricyclics, it increases norepinephrine levels. In higher doses, amantadine enhances dopamine transmission, both by increasing its release and blocking its reuptake. Structurally, amantadine is classified as an adamantane, as is memantine, a medication approved for Alzheimer's dementia that shares many of amantadine's pharmacodynamic properties.

How to Use Amantadine

Amantadine is dosed twice a day, usually in the morning and noon, with the bulk of the dose given in the morning to prevent insomnia. Its half-life is 16 hours. Amantadine is renally cleared and has no significant drug interactions. It can be added to other psychotropics or used as monotherapy. Most patients respond best to a daily dose of 200 mg, although some require 300 mg, and studies go as high as 400 mg. An extended release is available, but prohibitively expensive.

When introducing amantadine for depression:

- Start with 100 mg in the morning for one week

- If tolerated, increase to 100 mg twice daily (morning and noon)
- It may take up to three months to see full response
- Consider dose adjustment up to 300–400 mg/day for partial responders
- Continue for at least three to six months in full responders before considering discontinuation

Side Effects

Tolerability

Common side effects include insomnia, dizziness, nausea, and dry mouth. Although it has numerous psychiatric benefits, paradoxical reactions can also happen, such as when patients become more depressed, tired, or irritable on it.

Risks

Rarely, amantadine can cause delirium and hallucinations. I have seen hallucinations in three patients. For one, they were frightening, but the other two experienced them as therapeutic. They had a vision of Jesus Christ after one to two months on the drug, prompting them to take a more active role in church. Before amantadine, both were on disability for chronic depression that did not respond to dozens of medication trials over 10–20 years. After the vision on amantadine, both had full, sustained recovery. These cases are dramatic but rare and resemble reports of spiritual experiences on another glutamatergic drug, ketamine.

Amantadine has been associated with rare reports of hedonic homeostatic dysregulation (eg, compulsive gambling). On the cardiac side, amantadine can cause edema and orthostatic hypotension, which can lead to falls.

Other considerations before starting amantadine include:

- Baseline kidney function (dose adjustment needed in renal impairment)
- History of seizures or hallucinations (relative contraindications)
- Congestive heart failure (increased risk of peripheral edema)
- Cognitive status (although it can improve cognition, you need to monitor for delirium, especially in older adults)

TABLE 24-1. Amantadine Summary

FDA Approval	Dyskinesia in Parkinson's disease (as Gocovri XR), influenza A (as Symmetrel, instant release)
Off-Label Benefits	Depression, OCD, irritability (due to autism or traumatic brain injury), ADHD, post-COVID fatigue, tardive dyskinesia, weight loss, sexual dysfunction
Dosing	Start 100 mg Qam × 1 week then 100 mg BID (max 400 mg/day)
Risks	Edema, orthostasis, anticholinergic effects, skin pigmentation (livedo reticularis) Hallucinations, confusion, ataxia, hedonic dysregulation (impulsive gambling)
Side Effects	Insomnia, dizziness, nausea, dry mouth
Half-Life	10–25 hours
Interactions	Renally cleared (drugs that impair renal clearance may raise levels)

Key Takeaways

- Amantadine is a glutamatergic medication that may treat difficult-to-treat depression through effects on glutamate, norepinephrine, and dopamine.
- It may also treat common comorbidities of depression (OCD, ADHD, irritability, post-COVID fatigue, tardive dyskinesia, weight loss, and sexual dysfunction).
- Monitor for rare but serious adverse effects including hallucinations, confusion, and compulsive behaviors.

CHAPTER 25

D-Cycloserine

IN THE 1950s, PHYSICIANS NOTICED an unexpected side effect while testing two antibiotics in tuberculosis. The patients were more active, slept better, and some felt euphoric. One of those antibiotics went on to become the first antidepressant, the monoamine oxidase inhibitor (MAOI) iproniazid. The other was d-cycloserine (Seromycin).

Working at Montefiore Hospital in the Bronx, George Crane tested d-cycloserine in 67 psychiatric patients who also had tuberculosis. Half the patients had significant improvements in depression, anxiety, and sleep, but a few developed psychosis and mania. Those neuropsychiatric side effects limited its use, both in psychiatry and as an antibiotic, with frequent reports of psychosis, seizures, dizziness, slurred speech, and tremor. As pulmonologists turned to safer treatments for tuberculosis, psychiatric explorations of d-cycloserine receded to the animal labs, where we learned that it blocks glutamate.

Evidence in Depression

Trials of d-cycloserine in depression picked back up in 2006, spurred by the success of ketamine, another glutamate antagonist. But d-cycloserine only works as a glutamate antagonist in the high-dose range (above 500–750 mg/day). At lower doses it has the opposite effect, functioning as a glutamate agonist. That is likely why high-dose d-cycloserine (1,000 mg/day) worked as an augmentation strategy in treatment-resistant depression while low-dose (250 mg/day) did not. However, these promising results rest on only two small placebo-controlled trials, making any conclusions uncertain (Heresco-Levy U et al, *Int J Neuropsychopharmacol* 2013;16(3):501–506).

Learning from Low-Dose D-Cycloserine

Although low doses failed in acute depression, they do have a benefit that is proving useful here. They enhance learning. In animal models, low doses speed up extinction learning, helping the animal to adapt to a new environment without excessive anxiety. Similarly, low doses (50–150 mg) help patients overcome anxiety faster when taken one to two hours before an exposure exercise in psychotherapy (Rosenfield D et al, *J Anxiety Disord* 2019;68:102149). More than two dozen trials have tested this approach in phobias, social anxiety, post-traumatic stress disorder (PTSD), and obsessive-compulsive disorder (OCD), and research is exploring this augmentation strategy in transcranial magnetic stimulation (TMS).

D-Cycloserine in TMS

Exposure therapy and TMS both rely on neuroplasticity as a mechanism of change. Preclinical studies suggest that low-dose d-cycloserine enhances the neuroplastic effects of TMS, and a randomized trial tested this clinically in 50 patients. The benefit was large, with remission rates 10 times higher (39% vs 4%) in those who took d-cycloserine 100 mg one to two hours before each TMS session than those who took a placebo. D-cycloserine was only used during the first two weeks of the therapy, which was delivered as a four-week course of high intensity TMS (intermittent theta-burst stimulation) (Cole J et al, *JAMA Psychiatry* 2022;79(12):1153–1161).

Prevention after Ketamine

A few studies have tested d-cycloserine as a maintenance strategy after successful ketamine therapy, based on their shared glutamatergic mechanism. The results are mixed. Where it worked was in an industry-sponsored trial of bipolar depression, and this study used it in combination with lurasidone to uphold ketamine's benefits (Nierenberg A et al, *Int J Bipolar Disord* 2023;11(1):28). The pill, which combines 950 mg d-cycloserine with 66 mg lurasidone, is under development for bipolar depression, PTSD, and chronic pain by NeuroRx.

When to Consider D-Cycloserine

With its significant risks and limited empirical support, high-dose d-cycloserine is best reserved as a last resort for difficult-to-treat depression. Consider it in patients who:

- Did not respond to multiple conventional treatments
- Responded uniquely to glutamatergic medications like ketamine or amantadine but are not able to continue them
- Require maintenance therapy after successful ketamine treatment

There is better evidence to support the use of low-dose d-cycloserine, and there are lower risks when it is used to enhance:

- TMS therapy for treatment-resistant depression
- Exposure-based psychotherapy for comorbid anxiety disorders

Mechanism of Action

D-Cycloserine works as a glutamate agonist at low doses (50–150 mg/day) and becomes an antagonist at doses above 500–750 mg/day. This dose-dependent mechanism explains its different applications:

- At low doses, it enhances neuroplasticity and learning by modulating the N-methyl-d-aspartate (NMDA) receptor
- At high doses, it blocks excessive glutamate activity, similar to other glutamatergic medications like ketamine

How to Use D-Cycloserine

The availability of d-cycloserine is limited, so your patient may need to fill the prescription at an independent pharmacy that can take the extra steps to order it. It is only available as 250 mg capsules, which, although generic, are expensive ($10–$30 per capsule).

For Depression Treatment (the High-Dose Protocol)

- Start with 250 mg once daily
- Raise by 250 mg every five to seven days
- Target dose: 500 mg twice daily (1,000 mg/day total)
- Consider adding vitamin B6 50–200 mg/day to reduce neurologic side effects
- Assess response after four to six weeks at the target dose

For TMS Enhancement (the Low-Dose Protocol)

- Take 100 mg one to two hours before each TMS session
- Use only during the first two weeks of TMS treatment

TABLE 25-1. D-Cycloserine Summary

FDA Approval	Multi-drug-resistant tuberculosis
Off-Label Benefits	Depression Augmentation of TMS and exposure therapy
Dosing	Depression: 500 mg BID Augment TMS: 100 mg 1–2 hours before TMS for first 2 weeks of therapy Augment exposure therapy: 50–100 mg 1–2 hours before exposure session
Risks	In high doses: seizures (4%), psychosis, mania
Side Effects	Dizziness, headache, visual changes, tinnitus, tremor, confusion, drowsiness, slurred speech
Half-Life	10 hours
Interactions	Renally cleared (drugs that impair renal clearance may raise levels)

- A compounding pharmacy can repackage the 250 mg capsules into 100 mg sizes (for two weeks of TMS, the medication will cost $100–200)

For Exposure Therapy Enhancement

- Take 50–100 mg one to two hours before each exposure session
- No titration is needed

Side Effects

Tolerability

D-Cycloserine is well tolerated at low doses. At high doses, the most common side effects are dizziness (14%), headache (12%), visual changes (4%), tinnitus (5%), tremor, and drowsiness. Neurologic side effects are improved by taking it with vitamin B6 (50–200 mg/day).

Risks

Like other glutamatergic medications, high-dose d-cycloserine can cause psychosis. This risk did not show up in the depression trials, but occurs in 3% of patients who take it for tuberculosis. Other risks include seizures (4%), mania, and confusion.

Before prescribing high-dose d-cycloserine, consider:

- History of psychosis or mania (relative contraindication)

- Seizure history (relative contraindication)
- Cognitive status (monitor for confusion)
- Medication access and affordability

Low doses are safer and typically do not require special precautions. The low-dose protocol used in TMS is more affordable, as it only requires a compounding pharmacy to repackage six capsules.

Key Takeaways

- D-cycloserine has a dose-dependent effect on glutamate. At low doses, it enhances learning and functions as a glutamate agonist (50–150 mg); at high doses it treats depression as a glutamate antagonist (>500 mg).
- There is limited evidence that high-dose d-cycloserine (1,000 mg/day) is effective in difficult-to-treat depression; it should be considered only after failure of multiple conventional approaches.
- There is better evidence that low-dose d-cycloserine is effective when used to enhance other treatments, particularly TMS (100 mg before sessions) and exposure therapy for anxiety disorders.

CHAPTER 26

Minocycline

MINOCYCLINE (MINOCIN, DYNACIN) is a broad spectrum, tetracycline antibiotic that has been used to treat chlamydia, rickettsia, and mycoplasma pneumonia since 1971. In addition to blocking bacterial reproduction, the drug possesses other actions that suggested antidepressant potential. It is glutamatergic, neuroprotective, and has another property that points to where minocycline has proven most useful: it is anti-inflammatory.

Evidence in Depression

The first controlled trial of minocycline in depression tested it in HIV+ patients, a population with elevated inflammation (Emadi-Kouchak H et al, *Int Clin Psychopharmacol* 2016;31(1):20–26). That trial was positive, but seven subsequent trials in general depression were inconsistent. Half were positive; half negative. As researchers scurried to explain the disappointing results, they found two trends. Minocycline worked in patients with higher levels of inflammation (hs-CRP ≥ 3 mg/L) and in those who took it for a longer, three-month trial (Al Jumaili W et al, *Prim Care Companion CNS Disord* 2023;25(5):22r03467). A large trial is underway that will test those secondary findings. Pending that, little is certain about minocycline's antidepressant effects.

When to Consider Minocycline

Minocycline is best reserved as a last resort for patients with depression and an elevated high-sensitivity C-reactive protein (hs-CRP) ≥ 3 mg/L. Minocycline also treats a common comorbidity of depression, acne vulgaris, both as an oral medication (100 mg/day) and an FDA-approved foam. Consider minocycline for:

- Patients with treatment-resistant depression and elevated inflammatory markers or comorbid inflammatory conditions
- Patients with acne and depression, including lithium-induced acne (dual benefit)

Mechanism of Action

Minocycline has anti-inflammatory, neuroprotective, and glutamatergic effects that may explain its neuropsychiatric benefits. It inhibits several enzymes involved in the inflammatory cascade. These include the COX–2 enzyme where celecoxib (Celebrex) acts, lipoprotein-associated phospholipase A2 (a marker for inflammation and cardiovascular risk), and enzymes that produce inflammatory cytokines and nitric oxide. It has neuroprotective effects in animal models of traumatic brain injury, stroke, Alzheimer's, Parkinson's disease, and multiple sclerosis. Minocycline reduces glutamate release in the brain and blocks glutamate's excitotoxic effects at the microglia.

How to Use Minocycline

When using minocycline for depression:

- Start at 100 mg once daily for one week
- If tolerated, increase to 100 mg twice daily
- Take with a probiotic to minimize gut microbiome disruption
- Allow up to 3 months to assess full effectiveness
- After six months of recovery, consider tapering off minocycline (over two to four weeks)

While the optimal duration of treatment is unclear, extended antibiotic use raises concerns about bacterial resistance and microbiome effects.

Side Effects

Tolerability

Minocycline is generally well tolerated. The most common side effects are diarrhea, headache, dizziness, and photosensitivity (recommend sunblock with an SPF of at least 30 during the summer months).

Compared to other augmentation strategies for difficult-to-treat depression, minocycline offers:

TABLE 26-1. Minocycline Summary

FDA Approval	Bacterial infections. Acne vulgaris and rosacea (as topical and oral XR preparations).
Off-Label Benefits	Depression, negative symptoms of schizophrenia, rheumatoid arthritis
Dose	100 mg BID
Risks	Hepatotoxicity, fungal infection, altered microbiome Avoid if pregnant or allergic to any tetracycline
Side Effects	Diarrhea, headache, dizziness, photosensitivity
Half-Life	11–23 hours
Interactions	May accentuate effects of anticoagulants. Absorption decreased by antacids containing aluminum, calcium, magnesium, or iron.

- No weight gain or sedation
- No sexual side effects
- No mania risk
- Potential added benefit for acne

Risks

Two medical risks of minocycline raise concerns:

- Hepatotoxicity (very rare)
- Antibiotic resistance

On a population level, widespread use of antibiotics like minocycline can increase bacterial resistance. In individual patients, minocycline may alter the gut microbiome, which in turn can raise the risk of depression. That problem can be reduced by using minocycline short-term (eg, for less than six months) and prescribing it with a probiotic (Chapter 40: Probiotics). To put that risk in perspective, keep in mind that most antidepressants and antipsychotics also alter the microbiome by killing off selective bacterial strains.

Minocycline does not seem to cause mania. It has been studied in bipolar depression, but those results are not clear enough to recommend it there.

Before prescribing minocycline, consider:

- Liver function (check baseline LFTs)
- History of tetracycline allergy (contraindication)
- Pregnancy status (contraindicated)

- Concurrent medications that might interact (antacids, iron, anticoagulants)
- Plan for probiotic supplementation to support gut health

Key Takeaways

- Minocycline is an antibiotic with anti-inflammatory, neuroprotective, and glutamatergic properties.
- Its efficacy is uncertain in depression but is strongest for depression with elevated inflammatory markers (hs-CRP ≥ 3 mg/L) or in the context of inflammatory conditions.
- A longer treatment duration (up to three months) may be needed to observe full antidepressant effects.
- Consider combining minocycline with a probiotic to mitigate potential negative effects on the gut microbiome.

Rapid-Acting Medications

CHAPTER 27

Overview of Rapid-Acting Medications

THERE ARE TWO REASONS TO TURN to rapid action. First, when patients experience intolerable distress that could lead to self-destructive behavior, hospitalization, or treatment dropout. Second, when life circumstances call for a timely recovery, such as when symptoms threaten important ties at work, school, or in relationships, including the postpartum period.

Rapid-acting therapies are not new, but they've newly entered the mainstream with FDA approvals of esketamine (Spravato), zuranolone (Zurzuvae), and the bupropion-dextromethorphan combination Auvelity. Besides treating depression, these medications are anxiolytic, bringing needed relief during a crisis.

Their benefits, however, are short-lived. There is no solid evidence that any of these rapid-acting therapies prevent depression. Most of the rapid-acting therapies in this book are controlled substances, raising concerns about tolerance, withdrawal, and long-term depressogenic effects. They are easy to start, but difficult to stop.

Two exceptions are pindolol and thyroid. These have no rewarding qualities but speed up the response when started with an antidepressant. We discussed thyroid in an earlier chapter as this strategy can both accelerate and augment antidepressants, while pindolol only serves as a short-term accelerator.

We don't know how to maintain the benefits of these rapid-acting therapies, but history hints at an answer. In most cultures, healing is a ritual. It is a unique event, tinged with magic and danger. The surgeon does not operate on just anyone. Patients are carefully screened. Everything is set up to avoid a repeat operation, including discharge instructions to further the healing. The shaman's ritual is also dispensed with instructions for living.

TABLE 27-1. Rapid-Acting Medications for Depression

Rapid-Acting Medications for Depression
• **Alprazolam:** A benzodiazepine with short-term benefits in depression • **Eszopiclone:** A sleep medication that accelerates antidepressant response • **Auvelity:** A pill that pairs bupropion with dextromethorphan, a rapid-acting glutamatergic • **Ketamine/Esketamine:** A glutamatergic, dissociative anesthetic with acute antidepressant effects • **Pindolol:** A beta-blocker that speeds selective serotonin reuptake inhibitor (SSRI) response • **Thyroid:** A hormone that can speed up antidepressant response and augment when antidepressants fail • **Zuranolone:** A neuroactive steroid for postpartum depression • **Psychedelics:** Emerging treatments whose powerful psychological effects can be harnessed in psychotherapy

It is a singular event, designed to disrupt past patterns and inspire positive change.

The chapters in this section offer detailed guidance on using each medication, but a fundamental principle applies across all of them. These rapid-acting interventions are most effective when presented as catalysts for change rather than standalone solutions. At McGill University, Dr. Kyle Greenway and colleagues developed the Montreal Model to achieve that effect. Although it was developed for ketamine-assisted therapy, it provides a valuable framework that can apply to all rapid-acting medications (see page 223).

In this approach, clinicians set specific behavioral goals with patients before initiating treatment. The medication is presented as a tool to jump-start change, not as the change agent itself. Patients demonstrate readiness through initial behavioral steps before starting medication. After the acute effects subside, ongoing psychotherapy and lifestyle modifications sustain progress. Any "booster" treatments are used sparingly and only to enhance renewed behavioral efforts.

When integrated with a psychotherapeutic approach, rapid-acting medications offer meaningful relief during the most challenging phases of depression. By treating these interventions with the gravity they deserve, we can help patients find rapid relief while building sustainable recovery.

CHAPTER 28

Alprazolam and the Benzodiazepines

TO UNDERSTAND THE PROMISE AND PERILS of rapid-acting antidepressants, benzodiazepines are a good place to start. In the short term, these medications treat depression with an efficacy that rivals that of antidepressants. But their long-term effects are different and may even contribute to depressive symptoms.

Among the benzodiazepines, alprazolam (Xanax) came the closest to earning an antidepressant status. Its structure resembles that of a tricyclic, and it treated depression in 21 short-term trials. When Upjohn Pharmaceuticals first filed for approval of alprazolam in 1969, they registered it as an antidepressant. The FDA initially granted the request but later withdrew it as awareness of the addictive qualities of benzodiazepines grew in the 1970s. When it was finally approved for panic disorder in 1981, the product information mentioned that "anxiety associated with depression is responsive to alprazolam." That line is still there, but it is buried by multiple warnings about benzodiazepines causing depression and suicidality that have accumulated over the years.

Both these stories are true. Alprazolam treats acute depression, but it can worsen depressive symptoms with long-term use. Alprazolam shares that long-term risk with other benzodiazepines, but whether its acute benefits are unique among its class is open for debate. It is, however, the best studied of the benzodiazepines in depression.

Evidence in Depression

When started with an antidepressant, benzodiazepines accelerate the response with a small effect size (0.25), according to 10 placebo-controlled trials. Multiple benzodiazepines have been tested in this way, but alprazolam has the best evidence. Besides accelerating antidepressants, alprazolam is also effective as monotherapy in depression, at least over the short-term (four to six weeks). As monotherapy, it surpassed placebo and performed similarly to traditional antidepressants (selective serotonin reuptake inhibitors [SSRIs] and tricyclics), although with greater speed and better tolerability (from a Cochrane meta-analysis of 21 randomized controlled trials involving 2,693 subjects) (van Marwijk H et al, *Cochrane Database Syst Rev* 2012;2012(7)).

This efficacy is not just limited to anxious depression. Alprazolam treats core symptoms of depression (Benasi G et al, *Psychother Psychosom* 2018;87(2):65–74). However, there are significant concerns about its long-term use, which we'll discuss in the Risks section.

When to Consider a Benzodiazepine

Consider a benzodiazepine as a short-term measure to speed the effects of another antidepressant if the following criteria are met:

- The patient is under age 60, not taking an opioid, and has no history of substance use disorders
- There is an urgent need for rapid relief, such as to prevent hospitalization or job loss
- A working alliance is secure and the patient is motivated to make behavioral changes while starting the benzodiazepine
- The patient understands that the benzodiazepine is for short-term use

That last one requires some resolve but is not impossible. In a large study of commercially insured patients with depression, a small minority started a benzodiazepine with an antidepressant (11%). Among them, more than half (64%) came off the benzodiazepine after the first month and 88% came off within a year (Bushnell GA et al, *JAMA Psychiatry* 2017;74(7):747–755).

Even when those criteria are met, it may be preferable to use an FDA-approved option with rapid onset, such as esketamine, dextromethorphan-bupropion combo, or—in postpartum depression—zuranolone. Alprazolam is a pharmacodynamic cousin of zuranolone. The two act on the same receptor ($GABA_A$) and have similar sedating, anxiolytic, and rewarding qualities.

Mechanism of Action

Among the benzodiazepines, alprazolam has a unique structure that bears some resemblance to a tricyclic antidepressant; it has a triazole ring fused to its diazepine ring, a property it shares with another benzodiazepine, triazolam (Halcion). When stressed-out rats are given alprazolam, they become more active and their dopamine and serotonin transmission is enhanced. This effect is not seen with lorazepam (Ativan), and while it may contribute to alprazolam's antidepressant properties, it also suggests a risk. Increases in striatal dopamine like this are associated with a higher potential for addiction (Bentué-Ferrer D et al, *Eur Neuropsychopharmacol* 2001;11(1):41–50).

How to Use Benzodiazepines

Benzodiazepines are best limited to short-term use (two to six weeks) while waiting for another treatment to work or a crisis to pass. If addressing a crisis, treat it as a crisis. Meet at least weekly with the patient to assess their response to the medication and progress toward behavioral goals. Advise them to take the benzodiazepine only if they are out of bed and active during the day; in other words, if it helps them function.

Although alprazolam treats acute depression as monotherapy, it is usually taken with an antidepressant so that something is in place for long-term prevention. Some clinicians prefer to steer clear of alprazolam, as it has a reputation for diversion and misuse. Among the other options, clonazepam (Klonopin) (0.5–1 mg/day) has the next best evidence. Another way to reduce the misuse liability with alprazolam is to use the extended-release (XR) version, which is less rewarding than the instant release.

The average daily dose in depression is 3 mg (range 2–4 mg), but it is best to start low (0.5–1 mg/day) and raise as tolerated. Lower target doses are needed for patients who are frail or taking strong CYP3A4 inhibitors like fluvoxamine (Luvox) or nefazodone (Serzone).

After the 2- to 6-week course is complete, taper off the benzodiazepine over one to two weeks (eg, lower the dose by 50% over the first 4 days, then in smaller increments over the next 10 days).

Side Effects

Tolerability

Benzodiazepines are well tolerated in the short term. Drowsiness is the main side effect and usually improves within the first two weeks.

Risks

The main risk with benzodiazepines is a potentially lethal respiratory suppression when they are taken with opioids. Benzodiazepines are rarely fatal in overdose on their own, but can be when taken with opioids or alcohol. This risk is greater with alprazolam, clonazepam, and diazepam (Valium), and less with lorazepam and oxazepam (Serax).

Benzodiazepines also raise the risk of car accidents, falls, and respiratory illnesses. Long-term use can impair cognition, but this does not progress to dementia despite early reports of an association. Rather, it seems that anxiety and insomnia are early signs of impending dementia, and these symptoms prompt a benzodiazepine script (Osler M and Jørgensen MB, *Am J Psychiatry* 2020;177(6):497–505).

Tolerance, withdrawal, and misuse liability are serious risks with this class, particularly with rapid-acting, high-potency benzodiazepines like alprazolam.

When it comes to long-term effects on depression, observational studies comparing patients on long-term benzodiazepines with matched controls show concerning results. Patients on long-term benzodiazepines are less physically active, have worse cognition, and are less able to solve problems. If they come off the benzodiazepine, they have less depression and better cognition 6–12 months later (Allary A et al, *Aging Ment Health* 2024;28(12):1625–1633; Crowe SF and Stranks EK, *Arch Clin Neuropsychol* 2018;33(7):901–911; Barker MJ et al, *Arch Clin Neuropsychol* 2004;19(3):437–454).

TABLE 28-1. Alprazolam Summary

FDA Approval	Panic disorder Generalized anxiety disorder (GAD)
Off-Label Benefits	Acute depression
Dosing	0.5–3 mg XR QD for 4–6 weeks followed by a 2-week taper
Risks	Overdose when used with opioids Tolerance, withdrawal, misuse liability Falls, motor vehicle accidents Worsening of respiratory disorders
Side Effects	Drowsiness, cognitive dysfunction
Half-Life	6–27 hours
Interactions	Strong CYP3A4 inhibitors double levels (eg, grapefruit juice, fluvoxamine) while inducers lower them (eg, modafinil, carbamazepine)

Key Takeaways

- Benzodiazepines have fallen out of favor in depression, but their history as rapid-acting treatments offers valuable lessons as newer, branded medications with similar profiles enter the market.
- Benzodiazepines bring rapid relief while waiting for an antidepressant to work, but depressive symptoms can worsen with long-term use.
- Among the benzodiazepines, alprazolam has the best evidence in depression, but this benzodiazepine also has more rewarding qualities, raising concerns about addiction and misuse.

CHAPTER 29

Eszopiclone

TROUBLE SLEEPING IS ONE OF THE TOP predictors of depression, and restoration of sleep is a reliable indicator of recovery. So it was with great anticipation that researchers tested whether adding a sleep medication could speed up antidepressant response, and it was a great surprise when they failed. These were large, industry-sponsored trials, and with one exception, they failed to detect a benefit. That exception is eszopiclone* (Lunesta).

Evidence in Depression

Eszopiclone speeded up antidepressant response and brought about greater improvement in depression after one to two months in two large, placebo-controlled trials (Fava M et al, *Biol Psychiatry* 2006;59(11):1052–1060). Those trials were repeated with similar results in generalized anxiety disorder (GAD) (Pollack M et al, *Arch Gen* Psychiatry 2008;65(5):551–562). All of these studies enrolled patients with insomnia, but the results held up even when the investigators removed the sleep items from the rating scales.

However, other hypnotics did not pass this test. Not zolpidem (Ambien), and not the orexin antagonists (Schmidt ME et al, *Eur Neuropsychopharmacol* 2025;95:14–23). The reason may lie in a little-known secret of eszopiclone's pharmacology.

When to Consider Eszopiclone

Consider short-term eszopiclone to speed up and augment response in patients with major depressive disorder (MDD) or GAD who also have insomnia, particularly if rapid symptom relief is needed.

---

*Clinicians outside the US know eszopiclone through its racemic form, zopiclone. The two have similar properties, except zopiclone is dosed twice as high to account for the inert enantiomer (3 mg eszopiclone = 6 mg zopiclone).

Eszopiclone is not a good choice for the long-term management of depression or anxiety. Rather, it serves as a short-term intervention to accelerate response while waiting for the primary treatment to take full effect.

Mechanism of Action

Eszopiclone's edge in depression and anxiety may be due to its resemblance to benzodiazepines. Compared to other z-hypnotics, eszopiclone has broader effects on the $GABA_A$ receptor, attaching to the alpha–1 subunit involved in sleep as well as the alpha–2, -3, and -5 subunits involved in anxiety. Eszopiclone also produces an active metabolite with benzodiazepine-like and glutamatergic properties: desmethyl-zopiclone (Fleck MW, *J Pharmacol Exp Ther* 2002;302(2):612–618). This metabolite is anxiolytic but fortunately nonsedating, as it lingers during the day after taking eszopiclone.

The clinical data supports these pharmacodynamic similarities. Eszopiclone's anxiolytic effects are similar to those of benzodiazepines in the few studies that have compared them head-to-head.

How to Use Eszopiclone

Explain that eszopiclone is intended for short-term use to reset the sleep cycle and quickly reduce depression and anxiety. Once those symptoms have improved and sleep is stabilized, taper eszopiclone off over one to two weeks.

Start with 1 mg at bedtime, particularly in elderly patients or those with hepatic impairment. Increase to 2–3 mg as needed and tolerated. The full 3 mg dose may provide greater antidepressant benefits but also carries increased risk of side effects.

Eszopiclone should be taken immediately before bedtime, and patients should ensure they have a full 7–8 hours available for sleep to minimize morning hangover effects. Take without food or after a light snack; high-fat meals delay its absorption.

In the depression trials, eszopiclone was secretly replaced with a placebo after one to two months with no worsening of insomnia. This suggests that a four-to-eight-week course may be sufficient for most patients. Pair eszopiclone with behavioral strategies for sleep, such as regular wake times, caffeine reduction, a wind-down routine at night, and CBT-insomnia.

TABLE 29-1. Eszopiclone Summary

FDA Approval	Primary insomnia
Off-Label Benefits	Acute depression or generalized anxiety disorder
Dosing	1–3 mg QHS for 4–8 weeks, followed by 1–2 week taper
Risks	Sleep behavior disorders Falls Low risk of tolerance, withdrawal, misuse
Side Effects	Drowsiness, metallic taste, headache, dizziness
Half-Life	6 hours
Interactions	Strong CYP3A4 inhibitors double levels (eg, grapefruit juice, fluvoxamine) while inducers lower them (eg, modafinil, carbamazepine)

When discontinuing, taper gradually over one to two weeks to avoid rebound insomnia. This can be accomplished by alternating full doses with half doses for several days, then moving to half doses daily before stopping completely.

Side Effects

Tolerability

The main side effect is morning drowsiness, which improves by allowing 7–8 hours for sleep. Overall, eszopiclone's tolerability is similar to that of other z-hypnotics, though some patients complain of a metallic taste, which improves by taking it with a sip of a citrus drink (Yoshida M et al, *Chem Pharm Bull (Tokyo)* 2019;67(5):404–409).

Other common side effects include headache, dizziness, and dry mouth. These tend to be mild and transient, diminishing after the first several days of use.

Risks

Use of z-hypnotics like eszopiclone comes with a risk of REM sleep behavior disorders like sleepwalking, sleep eating, and sleep driving. Falls are also a risk, particularly in the elderly. Patients should be warned about these possibilities and advised to secure their sleep environment.

There's a small but significant risk of next-day impairment, particularly with the 3 mg dose or when sleep time is shortened. Patients should be cautioned about driving or operating machinery the morning after taking eszopiclone until they know how it affects them.

While eszopiclone has a lower misuse potential than benzodiazepines, it is still classified as a Schedule IV controlled substance. Prescribe with caution in patients with a history of substance use disorders, and limit the quantity dispensed during the initial treatment period.

Key Takeaways

- Short-term therapy with eszopiclone speeds antidepressant response in patients who have major depression or GAD along with insomnia.
- This effect is not seen with other sleep medications, possibly because eszopiclone's pharmacodynamic profile is closer to that of a benzodiazepine.
- Combine eszopiclone with behavioral methods for insomnia. Aim for a brief course (four to eight weeks) followed by a gradual taper.

CHAPTER 30

Auvelity

THE NAME "AUVELITY" CALLS TO MIND "augment with velocity," which is exactly what it does. This rapid-acting antidepressant is a combination of two older medications:

- Bupropion (Wellbutrin), an antidepressant from 1985, and
- Dextromethorphan (DM), a cough suppressant from 1958

Like ketamine, dextromethorphan is a glutamate antagonist. Their shared mechanism inspired early trials of dextromethorphan on its own in difficult-to-treat depression in the 2010s. From that success, Axome Therapeutics patented it as a combination pill with bupropion, which was approved as Auvelity in 2022.

Evidence in Depression

In eight early trials, dextromethorphan treated depression on its own, both in bipolar and unipolar disorders. But these trials were small and inconsistent, so it was not until the large industry-sponsored trials that we gained more confidence in its effects.

From those, we know that Auvelity starts to work after one week, faster than the two- to four-week wait for most antidepressants (Iosifescu DV et al, *J Clin Psychiatry* 2022;83(4):21m14345). It is also more effective than bupropion alone, with remission rates of 47% vs 16% at six weeks (Tabuteau H et al, *Am J Psychiatry* 2022;179(7):490–499). However, that efficacy edge did not translate into any benefit in an unpublished trial of treatment-resistant depression, a testament to how difficult this condition is to treat with medication.

When to Consider Auvelity

Auvelity's speed of onset makes it ideal for patients who are hospitalized or otherwise in need of rapid relief from depression. It is also worth considering as an alternative to the ketamines for patients who need oral, at-home administration.

Auvelity should be used cautiously in patients with a history of substance use disorders or seizures, as well as patients with bulimia (who have an increased risk of seizures on bupropion).

Mechanism of Action

Like ketamine, dextromethorphan (DM) inhibits glutamate transmission through N-methyl-d-aspartate (NMDA) antagonism. In addition, DM blocks serotonin and dopamine reuptake, complementing bupropion's effects on norepinephrine and dopamine. It also has anticonvulsant and neuroprotective effects.

The pairing with bupropion also accomplishes a pharmacokinetic goal by increasing DM's half-life. Bupropion inhibits DM's major metabolic pathway (CYP2D6), extending its half-life from 4 hours to 22 hours. A similar effect is achieved with Nuedexta, another branded formulation of DM that is FDA approved in pseudobulbar affect, a neurologic disorder characterized by sudden, uncontrollable bouts of laughter or tears. Nuedexta pairs DM with quinidine, a strong CYP2D6 inhibitor that otherwise adds nothing to Nuedexta's therapeutic effects.

How to Use Auvelity

The manufacturer recommends starting Auvelity at 45 mg DM/105 mg bupropion once daily for the first three days, then increasing to the target dose of 45 mg DM/105 mg bupropion twice daily. This gradual titration helps minimize side effects, particularly nausea, dizziness, and insomnia.

If the standard regimen is not tolerated, the two ingredients can be titrated separately, with DM given as a single dose in the evening to prevent daytime sedation. DM is available on its own as a prescription capsule or an over-the-counter syrup. The main advantage of the branded pill is that it lowers the risk of misuse and diversion, a risk that is likely higher with the over-the-counter syrup than the prescription DM capsule (Kwan ATH and McIntyre RS, *CNS Spectr* October 23, 2024).

In bipolar depression, DM is generally used without an antidepressant. Instead, it is paired with quinidine to extend its half-life (as DM 20–90 mg/day with quinidine 10 mg/day*).

Auvelity should be combined with a psychotherapeutic approach like the one described in the introduction to this section. With that in place, you can attempt to taper off DM while continuing the bupropion or other medications after one to three months of recovery. Withdrawal problems are possible but understudied, so taper slowly over two to six weeks. Patients may need to raise their bupropion to 300 mg/day during the withdrawal.

Side effects

Tolerability

Auvelity is associated with more side effects than bupropion, largely due to DM's sedative effects. One in 16 patients stopped Auvelity because of side effects, most often somnolence, nausea, dizziness, headache, and dry mouth.

Risks

Although Auvelity is not classified as a controlled substance, DM has a recognized misuse potential as "poor man's PCP," also known as "robotripping." There has been no evidence of inappropriate use in the clinical trials or postmarketing data, but patients with a recent history of substance use disorders were excluded from those trials.

The DM in Auvelity is dosed at a level (90 mg), well below those seen in recreational use (300–3,000 mg). However, the interaction with bupropion is likely to push DM's serum levels into the abusable range. The bupropion interaction raises DM's peak levels 40-fold higher and increases the total exposure to the drug 60-fold (ie, area under the curve) (O'Gvorman C et al, 2018 Poster Presentation ASCP Annual Meeting, Miami, FL).

DM can also cause psychosis, dissociation, and serotonin syndrome. These were not seen in the clinical trials of Auvelity, but they were reported in earlier investigations of DM's antidepressant effects.

---

*The 10 mg dose of quinidine has been discontinued, and it is only available in high doses (200 mg) that come with cardiac risks. However, this dosing can be achieved by combining a single tablet of Nuedexta (containing 10 mg quinidine with 20 mg DM) and adding extra DM if needed.

While case reports suggest DM can precipitate mania, small studies in bipolar depression found either no manic switching or improvement in manic symptoms (these studies paired DM with quinidine, not bupropion) (Lee SY et al, *Int J Bipolar Disord* 2020;8:11).

TABLE 30-1. Auvelity Summary

FDA Approval	Major depressive disorder Dextromethorphan (DM) is approved as a cough suppressant and (with quinidine, as Nuedexta) in pseudobulbar affect.
Off-Label Benefits	DM has evidence in bipolar depression and agitation in dementia.
Dosing	Start 45 mg/105 mg daily for 3 days, then 45 mg/105 mg BID Alternative: Titrate components separately with DM in evening
Risks	Bupropion: Seizures DM: Dissociation, psychosis, DM misuse
Side Effects	Drowsiness, nausea, dizziness, headache, dry mouth
Half-Life	2–4 hours (DM); 21 hours (bupropion)
Interactions	DM can cause serotonin syndrome with serotonergics and monoamine oxidase inhibitors (MAOIs); levels are raised and half-life extended by CYP2D6 inhibitors (a desirable interaction). Bupropion inhibits CYP2D6.

Key Takeaways

- Auvelity combines the ketamine-like dextromethorphan with the antidepressant bupropion to create a rapid-acting oral antidepressant that works within one week.
- Auvelity is not a controlled substance, but dextromethorphan has a misuse potential.
- Although Auvelity accelerates antidepressant response, it failed in an industry-sponsored trial of treatment-resistant depression.

CHAPTER 31

The Ketamines

KETAMINE IS A DISSOCIATIVE ANESTHETIC that was repurposed as a rapid-acting antidepressant in the early 2000s. It treats refractory depression with a large effect size, but two things have held it back. Ketamine's poor absorption requires IV delivery, and its generic status blocks its commercial path to FDA approval.

To get around the patent problem, companies have patented its mirror-image enantiomers, esketamine and arketamine. Esketamine earned FDA approval for treatment-resistant depression as Spravato in 2019, while arketamine is still being developed. Esketamine's original approval was as antidepressant augmentation. In 2025, the approval was updated to allow esketamine as monotherapy after a large trial demonstrated its efficacy without an antidepressant (unpublished trial, NCT04599855).

Esketamine is given intranasally because, like its parent compound, it is poorly absorbed in the GI tract. In this chapter, I will use "ketamine" when discussing both compounds and "IN esketamine" or "IV ketamine" when referring to them individually.

Evidence in Depression

Ketamine works quickly (within hours), but it's not clear how long its benefits last. The placebo-controlled trials are largely limited to the short-term, and in the case of esketamine the benefits diminshed over four weeks. The majority of IN esketamine's trials were positive in the first week but not at the one-month endpoint (Fountoulakis KN et al, *Am J Psychiatry* 2025;182(3):259–275).

Esketamine or Ketamine?

There's a general trend in psychiatry that intravenous medications are more effective than other routes of delivery, and this appears true for ketamine.

IV ketamine has a large effect size in treatment-resistant depression, compared to a small one for IN esketamine (Bahji A et al, *Expert Opin Drug Saf* 2022;21(6):853–866). It also has fewer negative trials.

Judging from head-to-head trials, IV ketamine is about as effective as ECT, though ECT has a clear advantage in psychotic depression, severe depression, and patients requiring hospitalization (ECT is also better at treating catatonia) (Jha MK et al, *JAMA Netw Open* 2024;7(6):e2417786). IN esketamine has never been compared to ECT but did surpass quetiapine (Seroquel) by a narrow margin in an eight-month randomized trial where both were used as augmentation in treatment-resistant cases (sponsored by the manufacturer of esketamine) (Reif A et al, *N Engl J Med* 2023;389(14):1298–1309).

Ketamine and Suicide

Ketamine is useful for acute crises, but does it prevent suicide? We can't say whether either form prevents suicide attempts, but IV ketamine or IN esketamine therapy prevents suicide, but IV ketamine does reduce acute suicidal ideation. Paradoxically, IN esketamine does neither, although it is FDA approved for "Depressive symptoms in adults with major depressive disorder with acute suicidal ideation or behavior." The language on that approval is confusing, but a close read of the package insert reveals that while IN esketamine reduced depressive symptoms in patients with suicidality, it did not reduce suicidality.

A troubling—but unsettled—controversy is whether the risk of suicide goes up after stopping ketamine. Four completed suicides and two suspicious deaths occurred in the FDA registration trials for IN esketamine. All occurred in the esketamine group, and most took place one to three weeks after the final dose. The FDA did not require a warning about this because the suicide rate didn't reach statistical significance.

Ketamine affects the opioid system, and opioid withdrawal raises the suicide risk (Oliva EM et al, *BMJ* 2020;368:m283). Studies also suggest a higher risk of suicidality during withdrawal from lithium, benzodiazepines, and possibly the serotonergic antidepressants (Maust DT et al, *JAMA Netw Open* 2023;6(12):e2348557; Moncrieff J et al, *J Aff Dis Rep* 2024;16:100765).

When to Consider Ketamine

Ketamine is to depression as opioids are to pain. It is useful for rapid relief, but best avoided long-term. Consider ketamine in these clinical scenarios:

- When rapid relief is needed for severe symptoms or to prevent hospitalization
- Treatment-resistant depression that has failed multiple medications and/or psychotherapy

One of ketamine's unique advantages is that several factors that predict a poor response to antidepressants predict a more favorable response to ketamine (Medeiros GC et al, *Transl Psychiatry* 2024;14(1):481; Lima Constantino J et al, *Psychiatry Res* 2025;345:116355). These include:

- History of childhood trauma
- Personal or family history of alcohol use disorder
- High levels of anhedonia, insomnia, or anxiety

Factors associated with poorer ketamine response include:

- Inflammatory markers
- Melancholic features

Ketamine is controversial in psychotic depression because the drug can cause psychotic symptoms. Although psychotic patients were excluded from the major trials, at least 99 case reports have tested the drug in psychotic depression without worsening of psychosis.

Ketamine-Assisted Therapy: The Montreal Model

At McGill University, Kyle Greenway and colleagues developed the Montreal Model to combine IV ketamine with psychotherapy. During the infusion, ketamine is delivered with evocative music and patients receive instruction on meditation. "We want them to cultivate a sense of curiosity, to be open to their own experience during the infusion." The model presents behavior change as the primary treatment. Ketamine is just there to jump-start the change.

"The first thing that we do is set goals. 'Imagine that your depression improves with ketamine. What activities do you see yourself doing to stay well?' Sometimes patients are so focused on symptoms that they forget the basics, like daily routines, sleep hygiene, exercise, and healthy diet. We aim for at least three SMART goals, which stands for Specific, Measurable,

Actionable, Realistic, and set in Time. It may be, 'I'm gonna get out of bed every day at 9 am and take a shower.'"

To ensure success, Dr. Greenway does not start ketamine until the patient is ready to change. "I look to see that we've set some goals and seen some progress on them. After setting goals, I may follow up with a phone call to check on their progress. If they say, 'I just can't find the motivation,' I'll say, 'Maybe we shouldn't rush into this.'"

After completing six ketamine treatments over the course of a month, patients continue this behavioral psychotherapy. If the patient relapses, Dr. Greenway may give ketamine again, but he makes sure these boosters are used to enhance the behavioral work and not replace it. "If needed, we may give a single dose every six months or so to restart the psychological work, but we don't repeat the full course."

The Montreal Model resembles psychedelic-assisted therapy. In both approaches, the therapist helps the patient harness the unique psychological experience of the drug to create more lasting life changes. Anecdotally, IV ketamine is more likely to bring about these heightened psychological experiences than IN esketamine, possibly because it reaches higher serum levels.

There are other models for ketamine-assisted therapy, including cognitive behavioral therapy (CBT), which sustained ketamine's benefits in a small, randomized trial (Wilkinson ST et al, *Psychother Psychosom* 2021;90(5):318–327).

Mechanism of Action

Ketamine has both psychological and biological effects that may explain its antidepressant actions. The psychological effects are immediate. Patients often feel like they are watching themselves from the outside, free from the worries of the ego and the negative chatter of ruminative thoughts. On the other hand, this out-of-body experience can trigger panic in some individuals.

These psychological effects have a biological basis, which probably rests in the default mode network that regulates self-conscious and ruminative thoughts. Ketamine quiets that chatter and alters default mode connectivity, changes that last for about 10 days after treatment (Evans JW et al, *Biol Psychiatry* 2018;84(8):582–590).

At the receptor level, ketamine functions as a glutamate antagonist, blocking transmission at the N-methyl-d-aspartate glutamate (NMDA) receptor. Ketamine shares this mechanism with amantadine, memantine (Namenda), topiramate (Topamax), and high-dose d-cycloserine. Downstream, ketamine regulates synaptic plasticity, raising neuroprotective compounds like brain-derived neurotrophic factor (BDNF), eukaryotic elongation factor 2 (eEF2), and glycogen synthase kinase–3 (GSK–3).

Ketamine also acts through the opioid system. In the prefrontal cortex, ketamine increases β-endorphins and revs up expression of the μ-opioid receptor gene. The opioid antagonist naltrexone (Revia) blocks ketamine's antidepressant effects (Jiang C et al, *Transl Psychiatry* 2024;14(1):90).

Ketamine is neuroprotective when used in low doses for short periods, but prolonged use and high doses can be neurotoxic (Motamedi-Manesh A et al, *Naunyn Schmiedebergs Arch Pharmacol* 2025;398(7):8111–8124).

How to Use Ketamine

Both formulations require careful administration and monitoring protocols. Each has distinct dosing requirements but similar safety considerations.

Intranasal (IN) Esketamine Protocol

IN esketamine follows an FDA-approved protocol with three phases:

Dosing Schedule

- Induction phase: Twice weekly for four weeks
- Continuation phase: Once weekly for four weeks
- Maintenance phase: Every one to two weeks based on individual response

Dose

- Starting dose: 56 mg
- Target dose: 84 mg (increase as tolerated)
- Maximum FDA-approved dose: 84 mg

Administration Requirements

- Risk Evaluation and Mitigation Strategies (REMS) enrollment for prescribers and facilities

- Pre-administration assessment: Blood pressure, mental status
- Supervised administration in certified healthcare setting
- Post-administration monitoring for at least two hours
- Vital sign measurements at 40 minutes and two hours
- Assessment for sedation and dissociation before discharge
- Patient must be accompanied home and cannot drive on treatment day

Intravenous (IV) Ketamine Protocol

IV ketamine is not FDA-approved for use in depression, so protocols vary across treatment centers.

Typical Dosing Schedule

- Induction series: Twice weekly for two to three weeks (six infusions total)
- Maintenance: Variable, typically every two to four weeks based on symptoms

Dose

- Standard dose: 0.5 mg/kg administered over 40 minutes
- Some centers use 0.25–0.75 mg/kg based on response and tolerability

Although IV ketamine may be more effective than IN esketamine, its lack of FDA approval means that patients usually need to pay privately at a ketamine clinic. IN esketamine, in contrast, is usually covered by insurers after prior authorization. The choice between them often comes down to cost, but one form to avoid is oral ketamine. The oral form is not well studied, and its irregular absorption means it is less likely to work. It also opens the door to misuse and diversion.

Maintenance Treatment Strategies

Some patients remain well after the initial two-month course, but many relapse, and it's not clear what the best approach to prevention is. Cognitive behavioral therapy has some support but attempts to sustain the benefits with other medications have largely failed, including lithium as well as medications that share in ketamine's glutamatergic mechanism, like lamotrigine (Lamictal), riluzole (Rilutek), and d-cycloserine (McMullen EP et al, *Adv Ther* 2021;38(6):2795–2820).

Another option is to continue the drug at a reduced frequency, much as we do with maintenance electroconvulsive therapy (ECT). This was tested with IN esketamine in three long-term maintenance trials involving more than 2,000 patients followed for one to three years: SUSTAIN–1, –2, and –3. These trials, which were sponsored by the manufacturer, did not uncover any tolerance, psychosis, cognitive impairment, or new safety concerns.

Only one of these trials employed a controlled design, randomizing 297 patients to placebo or maintenance IN esketamine after responding to the drug. Over the next four months, relapse rates were much lower with maintenance IN esketamine (26%) than placebo (45%–59%) (Daly EJ, *JAMA Psychiatry* 2019;76(9):893–903).

In these trials, maintenance doses were given every one to two weeks. In practice, most patients stay at the two-week interval, but it is worth attempting to decrease the frequency, first to three then four weeks. This tends to go better if their life is going well or they have made significant gains in psychotherapy. Relapses can be addressed with a re-initiation of IN esketamine using the acute treatment protocol (Castro M et al, *CNS Drugs* 2023;37:715–723).

Medications to Avoid with Ketamine

Benzodiazepines

- May blunt ketamine's antidepressant effects.
- Ketamine helps manage benzodiazepine withdrawal.
- Consider tapering benzodiazepines over one to two months before starting ketamine.
- Plan final benzodiazepine dose one to two days before first ketamine treatment.
- This approach succeeds in 91% of patients attempting benzodiazepine discontinuation (Garel N et al, *Neuropsychopharmacology* 2023;48(12):1769–1777).

Stimulants

- The FDA warns about additive hypertensive effects when combined with ketamine.
- Potential for increased neurotoxicity and psychosis.
- Consider tapering off or temporarily discontinuing on treatment days.

Side Effects

Tolerability

Side effects to ketamine include sedation, dissociation, nausea, transient cognitive impairment, and (with IN esketamine) a bitter taste. Most of these are transient and resolve within a few hours of treatment.

Risks

Ketamine requires close monitoring for three acute risks:

- Hypertension (15%)
- Sedation (20%)
- Changes in mental status (illusions, space-time distortions, perceptual changes, derealization, and depersonalization)

These mental status changes are transient but common, occurring in 60%–80% of patients in the IN esketamine trials. In rare cases, the sedation can cause loss of consciousness or respiratory depression, which is part of the reason that the FDA requires in-person monitoring.

With high doses or prolonged use, potential adverse effects include:

- Bladder dysfunction (ulcerative or interstitial cystitis)—assess by asking about dysuria and hematuria
- Cognitive impairment from neurotoxicity
- Tolerance and dependence
- Conversion to recreational use (eg, patients may seek extra doses outside the clinic)
- Potential for mood worsening during withdrawal
- Psychosis

Psychosis is a particular concern given ketamine's use as a pharmacological model for schizophrenia. Ketamine worsens positive and negative symptoms when given to patients with schizophrenia, and it has similar effects when given to healthy controls (Beck K et al, *JAMA Netw Open* 2020;3(5):e204693). About half of chronic recreational users of ketamine develop cognitive and psychotic symptoms that closely resemble schizophrenia (Cheng WJ et al, *Schizophr Res* 2018;199:313–318; Luo T et al, *Front Psychiatry* 2022;12:786622).

On the other hand, there are only two case reports of sustained psychosis after ketamine therapy, and both had complicating factors (one was using cannabis, and the other had post-COVID depression) (Pacilio RM and Geller JA, *J Clin Psychopharmacol* 2025;45(3):290-291; Tang J et al, *Encephale* 2024;50(5):583–584).

TABLE 31-1. Ketamine Summary

FDA Approval	IN esketamine: Treatment-resistant depression, depression with acute suicidality IV ketamine: anesthetic (not approved in psychiatry)
Off-Label Benefits	Bipolar depression, post-traumatic stress disorder (PTSD)
Dosing	IN esketamine: 56–84 mg twice weekly for 4 weeks, then taper off or continue as weekly for 4 weeks, then every 1–2 weeks IV ketamine: 0.5 mg/kg over 40 minutes, typically twice weekly for 2–3 weeks
Risks	Hypertension, dissociation, psychosis, mania Bladder dysfunction with long-term use Possible increased suicide risk during withdrawal
Side Effects	Sedation, nausea, transient cognitive impairment, dizziness, bitter taste
Half-Life	2–4 hours
Interactions	Metabolized by CYP2B6 and CYP3A4 with minor contributions from CYP2C9 and CYP2C19 Naltrexone and benzodiazepines may interfere with response Additive hypertensive effects with stimulants

Key Takeaways

- Ketamine (IV) and esketamine (IN) provide rapid relief of depressive symptoms within hours to days, but we don't have good evidence that they work beyond a few weeks.
- Of the two, ketamine (IV) is likely more effective, but it lacks FDA approval and often requires out-of-pocket payment.
- Long-term use carries risks of bladder dysfunction, cognitive impairment, psychosis, conversion to recreational use, tolerance, and withdrawal.
- The Montreal Model prevents long-term use by combining ketamine with active behavioral therapy.

CHAPTER 32

Pindolol

PINDOLOL IS DIFFERENT FROM OTHER rapid-acting treatments. It is not a controlled substance, and it is not very rapid. It speeds up response to selective serotonin reuptake inhibitors (SSRIs) by a few weeks, so that patients respond after two weeks instead of three to six weeks.

Pindolol is a beta-blocker FDA-approved for hypertension. It also has serotonergic effects that suggest a synergy with serotonergic antidepressants.

Evidence in Depression

Studies of pindolol in depression began in the 1990s and number approximately 11 today. When meta-analyzed together, the results show that pindolol speeds up SSRI antidepressants, doubling the rate of response at two weeks. However, by four weeks the differences from placebo are negligible, and pindolol does not reliably augment antidepressants that have failed to work (Liu Y et al, *Hum Psychopharmacol* 2015;30(3):132–142; Portella MJ et al, *J Clin Psychiatry* 2011;72(7):962–969).

Pindolol only accelerates response to SSRI antidepressants, and it works best when the two are started at the same time.

Besides accelerating their antidepressant effects, pindolol may augment SSRIs in panic disorder and OCD, a tentative finding from small, placebo-controlled trials. In those disorders, pindolol worked as augmentation, not acceleration, at a dose of 2.5 mg TID (Hirschmann S et al, *J Clin Psychopharmacol* 2000;20(5):556–559; Sassano-Higgins SA and Pato MT, *J Pharmacol Pharmacother* 2015;6(1):36-38).

When to Consider Pindolol

Pindolol is worth considering when rapid response to an SSRI is needed but other options in this section are not appropriate, such as in patients at risk for substance misuse. Specific clinical scenarios where pindolol may be beneficial include:

- First-time SSRI treatment when rapid response is desirable
- When restarting an SSRI after a period off the drug
- When rapid antidepressant effect is needed, but other options in this book are contraindicated
- Patients on SSRIs with comorbid panic disorder or OCD

Pindolol is not indicated for patients with:

- Treatment-resistant depression
- Contraindications to beta-blockers (see Risks section)

Mechanism of Action

Pindolol treats hypertension through nonselective beta-receptor blockade, but its antidepressant actions involve serotonin. As a serotonin–1A receptor partial agonist, pindolol potentiates the initial rise in serotonin that SSRIs produce.

When a patient first takes an SSRI, there is an immediate increase in serotonin in the synaptic cleft. However, this activates inhibitory serotonin–1A autoreceptors in the raphe nucleus, which reduces serotonergic firing and limits the initial serotonergic effect. Pindolol blocks these inhibitory autoreceptors, preventing this negative feedback mechanism and allowing for greater serotonergic transmission from the start of treatment.

Pindolol's ability to bind at this receptor is diminished in patients who are already taking SSRIs, suggesting it is more likely to work in antidepressant-naïve patients than in those with treatment resistance. The clinical data backs that up.

How to Use Pindolol

Pindolol accelerates but does not improve antidepressant response, so short-term use is ideal. Start pindolol with an SSRI, and plan to taper off after four to six weeks, gradually lowering the dose over two to four weeks. Pindolol is often started at the treating dose (5 mg TID), although it can be titrated more slowly (eg, over a week) if needed. The target dose in panic disorder and OCD is lower (2.5 mg TID).

Side Effects

Tolerability

Pindolol is well tolerated. Insomnia, vivid dreams, and edema are possible.

Risks

Avoid pindolol in patients with heart disease. It is contraindicated in congestive heart failure, severe bradycardia, second- and third-degree heart block, and cardiogenic shock. It can also cause bronchial constriction so is contraindicated in bronchial asthma. Although pindolol does not generally have withdrawal symptoms, there are rare reports of myocardial infarction after abrupt discontinuation in patients with preexisting heart disease.

TABLE 32-1. Pindolol Summary

FDA Approval	Hypertension
Off-Label Benefits	Accelerating SSRI response Potential benefits in panic disorder and obsessive-compulsive disorder (OCD)
Dosing	5 mg TID Duration: 4–6 weeks followed by 2–4 week taper
Risks	Contraindicated in asthma, heart failure, severe bradycardia, heart block, cardiogenic shock
Side Effects	Insomnia, vivid dreams, fatigue, dizziness, peripheral edema
Half-Life	8 hours
Interactions	Metabolized through glucuronidation and hydroxylation. Hypotension with other beta-blockers. May reduce effectiveness of beta-agonist inhalers.
Monitoring	Blood pressure and heart rate at baseline and during dose changes

Key Takeaways

- Pindolol accelerates SSRI response by approximately two weeks but does not enhance overall efficacy.
- Benefits are limited to antidepressant-naïve patients; it is not effective in treatment-resistant depression.
- It must be started simultaneously with the SSRI to achieve accelerating effects.

CHAPTER 33

Zuranolone

ZURANOLONE (ZURZUVAE) IS A RAPID-ACTING medication approved in 2023 for postpartum depression (PPD). Rapid onset is particularly important in PPD, where every day of active symptoms takes a measurable toll on infant development. Zuranolone brings relief within 3 days.

Zuranolone is an oral version of IV brexanolone (Zulresso), which was approved for the same indication in 2019. Brexanolone was taken off the market in 2025 due to low sales, hindered by its high cost and requirement for overnight monitoring.

Evidence in Depression

Zuranolone demonstrated efficacy in postpartum depression, with a medium effect size (0.5) in randomized, placebo-controlled trials. Zuranolone worked quickly, within 3 days, with benefits persisting through the 14-day course (Deligiannidis KM et al, *JAMA Psychiatry* 2021;78(9):951–959). Its benefits declined a month after stopping it—the longest follow-up period—but remained superior to placebo.

Most of the women in the zuranolone trials were in their first episode of depression, although a small minority (14%) had recurrent depression. The medication was not given during pregnancy, and the women had to be within a year of childbirth to start it. For some, the depression began in the third trimester, although in most cases (63%) it started within the first four weeks of delivery. Those with suicidality, psychosis, bipolar disorder, or recent substance use disorders were excluded.

What about nonpostpartum depression? Zuranolone was tested there, with promising results in small, early trials, but diminishing returns in larger trials. The medication had rapid effects but they faded after a few days, leading the FDA to reject zuranolone for nonpostpartum depression (Carvalho T, *Nat Med* 2023;29(5):1032–1033).

When to Consider Zuranolone

Zuranolone is ideal for women with postpartum depression who:

- Have depression that began in the third trimester or within the first month after delivery
- Do not have recurrent depression, psychosis, or bipolar disorder
- Need rapid symptom relief

We don't know if zuranolone works in late-onset postpartum depression, a term used for depressions that begin 2–12 months after delivery. However, there is one late-onset type where zuranolone poses a risk: Women who develop depression after stopping breastfeeding. Allopregnanolone tends to rise when women wean from breastfeeding and resume menstruation. Paradoxically, this rise is linked to depression, suggesting an allopregnanolone analog like zuranolone could have undesirable effects in these cases (Burke CS et al, *Arch Womens Ment Health* 2019;22(1):55–63).

Mechanism of Action

Zuranolone is one of few psychiatric medications that can claim to treat an underlying cause. It is a synthetic version of allopregnanolone, a hormone that rises during pregnancy and takes a sharp fall after delivery.

Allopregnanolone modulates the $GABA_A$ receptor that is involved in anxiety, mood, and cognition. This is the same receptor that benzodiazepines attach to, so the analogy here is a bit like benzodiazepine withdrawal. Mood worsens when allopregnanolone falls, and improves when it is temporarily replenished by the medication. Postpartum depression is linked to changes in the GABAergic neurons, further supporting zuranolone's role here (Guintivano J et al, *Am J Psychiatry* 2023;180(12):884–895).

Like the benzodiazepines, zuranolone is a positive allosteric modulator of the $GABA_A$ receptor. Unlike the benzodiazepines, it binds to a different area of $GABA_A$, upregulates the receptor, and has broader effects on GABAergic transmission. Industry-sponsored papers have used this pharmacodynamic difference to imply that zuranolone will not cause tolerance like the benzodiazepines do, but the theoretical evidence for this is weak, and the clinical evidence is nil (Stahl SM et al, *CNS Spectrums* 2023;28(2):260–261).

How to Use Zuranolone

Zuranolone is dosed once at night for two weeks. It needs to be taken with a high-fat meal of at least 400 calories to be absorbed. The starting and treating dose is 50 mg, and this can be lowered to 30 mg if it causes problematic side effects like fatigue. Both the 30 mg and 50 mg doses brought similar benefits in the postpartum trials.

It can be taken with antidepressants (15%–20% of the women in the studies were taking them) but whether it will work with a GABAergic medication like a benzodiazepine or z-hypnotic on board is less clear; patients were not allowed to take those during the trials.

Maintaining the Benefits

Zuranolone's main advantage is its rapid onset. The drug is meant to be taken as a short-term two-week course, but what happens after those two weeks?

We don't know what prevents depression after this treatment. Reasonable strategies include antidepressants, lithium, exercise, sleep regulation, or psychotherapy. A repeat course of zuranolone is possible, but I wouldn't repeat it more than one or two times.

The FDA has not addressed the issue of repeated zuranolone trials, but this strategy was tested in a large open-label study of nonpostpartum major depression. In the year following treatment, 55% of patients required a second two-week course of zuranolone, and 16% required three or four additional courses (Meshkat S et al, *J Affect Disord* 2023;340:893–898).

Side Effects

Tolerability

Zuranolone's main side effects are drowsiness, dizziness, and diarrhea. One in four women experienced significant fatigue, although unlike brexanolone, it did not cause loss of consciousness.

Risks

Zuranolone is not recommended in breastfeeding women (its safety has not been tested in this population). This presents a dilemma, as women may lose the ability to breastfeed if they take a hiatus for the two-week treatment. Pumping milk during that period may keep lactation intact.

In addition to sharing a common mechanism with the benzodiazepines, zuranolone shares in their anxiolytic, hypnotic, and rewarding effects. One study asked subjects with a history of sedative misuse to compare zuranolone, alprazolam, and placebo. Zuranolone made them feel euphoric and drunk, and they found it just as rewarding as alprazolam, leading the FDA to classify it as a Schedule IV controlled drug along with the benzodiazepines.

Zuranolone impairs motor skills and comes with a warning not to drive for 12 hours after taking it.

TABLE 33-1. Zuranolone Summary

FDA Approval	Postpartum depression
Dosing	30–50 mg QHS for 2 weeks Take with a high-fat meal
Risks	Driving impairment, risk of misuse Do not use while breastfeeding
Side Effects	Sedation, diarrhea
Half-Life	16–23 hours
Interactions	Strong CYP3A4 inhibitors double levels (eg, grapefruit juice, fluvoxamine) while inducers lower them (eg, modafinil, carbamazepine)

Other Postpartum Disorders

Depression is not the only disorder with a postpartum onset. There are other syndromes to consider in this period, but first let's look at what defines "postpartum."

DSM sets the postpartum cutoff at 4 weeks after delivery; ICD-11 at 6 weeks, while other groups argue for 6 to 12 months. Actually, DSM-5 changed this definition to include cases that began during pregnancy and renamed it peripartum onset. The zuranolone trials enrolled women whose depression began in the third trimester (37%) although most had their onset in the first four weeks after delivery (63%).

Other postpartum disorders to rule out include:

1. Postpartum blues ("baby blues")

This is a mild syndrome of tearfulness, irritability, and insomnia. It starts 1 to 2 days after delivery and resolves within 10 days. Zuranolone is not appropriate here, but careful follow-up is. Women with postpartum blues are twice as likely to develop PPD (Landman A et al, *Eur Psychiatry* 2024;67(1):e30).

2. Postpartum Obsessive-Compulsive Disorder (OCD)

OCD is very common postpartum, affecting around 1 in 6 women (Fairbrother N et al, *J Clin Psychiatry* 2021; 82(2):20m13398). On the other hand, half of women with postpartum depression have obsessive thoughts, usually centered on the child's health and safety. Further complicating the distinctions, both OCD and postpartum onset are risk factors for bipolar disorder. While zuranolone does not treat OCD, it isn't known to make it worse. As long as the OCD symptoms occur alongside significant depression, zuranolone would be appropriate.

3. Postpartum psychosis

Postpartum psychosis is a psychiatric emergency that can threaten the life of mother and child. It comes on quickly, peaking at 10 days postpartum. These patients go in and out of lucidity and confusion, with waxing and waning delusions or hallucinations that usually center on fears of the baby's safety and paranoia toward their relatives. The majority have bipolar features, and treatment guidelines recommend lithium and antipsychotics first line and caution with antidepressants, which can worsen the manic side of this syndrome (Jairaj C et al, *Focus (Am Psychiatr Publ)* 2024; 22(1):131–142). Zuranolone does not treat psychosis.

4. Postpartum bipolar

Any postpartum episode should raise the index of suspicion for bipolar disorder, particularly if the symptoms are severe or involve psychosis. When postpartum episodes require hospitalization, the chance of a bipolar diagnosis rises eight-fold (Munk-Olsen T et al, *Arch Gen*

Psychiatry 2009;66(2):189–195). Zuranolone is not known to cause mania. It appeared safe in a small, phase II trial of bipolar depression, but whether it treats postpartum bipolar disorder is unknown.

5. Medical causes

Complications of pregnancy such as preeclampsia raise the risk of postpartum depression, but there are also disorders worth ruling out here, particularly low iron and hypothyroidism, which are more common postpartum. The postpartum period is also an inflammatory state. All of the treatments for inflammatory depression are worth considering, although only celecoxib has specific evidence there (Esalatmanesh S et al, *Arch Gynecol Obstet* 2024;309(4):1429–1439).

Key Takeaways

- Zuranolone is a rapid-acting neurosteroid for postpartum depression, but it does not work for other types of depression.
- It is a synthetic version of allopregnanolone, a hormone that normally rises during pregnancy and falls after delivery.
- It has rewarding qualities similar to benzodiazepines and is intended for short-term use.

CHAPTER 34

Psychedelics

AT THE TIME OF THIS WRITING, psychedelics await approval from the FDA, but many patients are not waiting. They find them through underground clinics, cannabis dispensaries, or abroad. In Oregon, they are legal, and psychedelic-assisted therapy occurs out in the open with minimal state regulation.

Psychedelics are drugs that alter consciousness and perception, and they are divided into two types. Classic psychedelics activate the serotonin 2A receptor. They include psilocybin, dimethyltryptamine (DMT), and lysergic acid diethylamide (LSD). Atypical psychedelics achieve similar psychological effects through different mechanisms. They include the amphetamine derivative MDMA (3,4-methylenedioxymethamphetamine) and the glutamatergic ketamine.

Whether this altered state of consciousness is necessary for their antidepressant effects is open to debate, as is the necessity of pairing them with psychotherapy. My own bias is that the context in which they are taken matters, but I believe that is true for most psychotropics.

Evidence in Depression

Of the half-dozen psychedelics under development, psilocybin is the closest to consideration in depression. Also known as "magic mushrooms," psilocybin has been used in religious ceremonies for at least 9,000 years. It was rediscovered by Western society in the 1950s, and for a few years was available as a prescription medication to speed the process of psychoanalytic therapy.

Psilocybin's first modern trials involved patients with terminal cancer, where a single dose relieved anxiety about death in 70% of patients for up to six months (Dodd S et al, *CNS Spectr* 2023;28(4):416-426). Studies in depression came next and have matured to the final phase III level.

In depression, psilocybin has positive results from two randomized controlled trials that tested the drug in the context of supportive psychotherapy. The first compared psilocybin to escitalopram (Lexapro) 20 mg/day in 59 patients with longstanding major depressive disorder (mean duration 22 years). Psilocybin passed, but did not live up to its hype, with no statistical differences between the two groups (Carhart-Harris R et al, *N Engl J Med* 2021;384(15):1402–1411).

The second trial tested psilocybin in a treatment-resistant population, after two to four antidepressant failures. This large trial compared two doses of psilocybin (25 mg vs 10 mg) to placebo. The results were positive, but not without disappointment. Psilocybin separated from placebo on the primary outcome measure at three weeks, but not at 12 weeks, and only the high dose (25 mg) was effective. This dose was also associated with a high rate of side effects, with 84% reporting headache, nausea, or dizziness (Goodwin GM et al, *N Engl J Med* 2022;387(18):1637–1648).

One problem with psychedelic research is that the studies attract people who had a positive past experience on the drug, and steer away those with the opposite. This problem is compounded by the fact that blinding is difficult in these trials. Over 90% of patients can tell if they took a psychedelic, and the researchers who observe them can usually tell as well. One in three participants in the first depression trial had tried psilocybin in the past, and a reanalysis of this trial confirmed that expectations biased the results (Dutcher EG and Krystal AD, *JAMA Psychiatry* 2025;82(3):321–322).

Who is Likely to Respond

We are ready to recommend psychedelics, but need to understand their potential to guide patients who seek them out on their own. Research suggests psilocybin may have potential benefits in:

- Treatment-resistant depression
- Existential anxiety about mortality in terminal illness

Psilocybin therapy is less appropriate for:

- Patients with bipolar or psychotic disorders or a family history of psychosis
- Those with serious cardiac conditions
- Patients taking serotonergic medications (due to risk of serotonin syndrome)

Mechanism of Action

Psilocybin activates the serotonin 5-HT2A receptor, operating like a less potent version of LSD. It tends to induce a self-transcendent state along with perceptual hallucinations that range from the wondrous to the frightening.

Psilocybin reduces depressive rumination and can bring about many unique psychological effects. After the treatment, people tend to be more open, spiritual, connected to nature, extraverted, altruistic, and less likely to endorse authoritarian political views. The psilocybin experience has a "noetic" quality, which means the person has a sense of profound truth that is difficult to put into words. Many subjects described it as the most meaningful experience of their lives.

One reason to pair psychedelics with psychotherapy is to extend their benefits. Psychedelics have rapid effects, inducing a more flexible, connected, self-transcendent state. These changes are temporary, lasting from one to two weeks with ketamine to over six to twelve months with psilocybin. Repeated dosing can lead to tolerance and other problems, so something else is needed. Psychotherapy is well suited to move patients from that newfound awareness to new ways of living, and that behavioral change can prevent depression.

Dosing Protocols in Trials

In clinical trials, psilocybin is given in one or two dosing sessions, each lasting four to eight hours. One or two psychotherapists are present to provide support, buffering the potentially frightening visions that psilocybin conjures up. These dosing sessions are followed by traditional psychotherapy sessions where patients integrate the new perspective they gained from the drug. Most trials used a version of acceptance and commitment therapy (ACT), which encourages clients to create a more meaningful life through purpose-driven action and a more flexible mindset.

The standard protocol includes:

- Preparatory sessions (one to three visits)
- Dosing session (25 mg psilocybin, administered orally)
- Integration sessions (three to four visits over one to two months)

If patients seek out psilocybin through nonmedical channels, advise them to:

- Start psychotherapy before dosing
- Take no more than two doses, and start with a low dose (eg, 10–15 mg)
- Have a trusted support person present during the experience
- Take it in a safe, comfortable environment
- Avoiding taking it with serotonergic antidepressants or other recreational drugs

Side Effects

Tolerability

The most common side effects to psilocybin are anxiety, headaches, nausea, confusion, vomiting, and increases in blood pressure and pulse. These side effects are usually transient and mild. In the clinical trials, there are no reports of serious side effects that required intervention (Johnson MW et al, *Neuropharmacology* 2018;142:143–166).

Risks

Psilocybin's lethal dose is approximately 1,000 times its therapeutic dose.

Negative feelings are part of most psilocybin trips, but most people report that the good eventually outweighs the bad. In an online survey of 1,993 people who experienced psychological difficulties after taking psilocybin, 39% said it was among the most challenging experiences of their life but most of them (84%) felt they benefited from the challenge. Anxiety, depression, and paranoia were common and lasted more than a week for 1 in 4 respondents, and more than a year for 1 in 10, and 8% sought professional help for these effects. More concerning were the 11% who put themselves or others in physical danger after taking the drug (Carbonaro TM et al, *J Psychopharmacol* 2016;30(12):1268–1278).

Hallucinations and illusions are common reactions to a psilocybin dose. None of these persisted in the clinical trials, but they can with recreational use. This is diagnosed in DSM-5 as Hallucinogen Persisting Perception Disorder and causes visual disturbances like dripping colors or flashbacks to earlier psychedelic trips at a rate of 4% in recreational users. Most psychedelics are linked to mania and psychosis, including new onset schizophrenia. However, it is not clear if the drugs caused these syndromes or hastened their onset in vulnerable people.

Psychedelics break down interpersonal boundaries. That raises the risk of boundary violations, a problem that has already transpired in the MDMA trials. The most egregious case began with cuddling during the dosing session and progressed to a sexual relationship, despite the use of video monitoring and dual therapists to prevent this kind of breach. The violation was one of the reasons that MDMA failed to gain FDA approval for post-traumatic stress disorder (PTSD).

TABLE 34-1. Psilocybin Summary

FDA Approval	None
Off-Label Benefits	Existential despair in terminal illness Major depression Potential benefits in substance use disorders, obsessive-compulsive disorder (OCD), and cluster headaches
Dosing	25 mg single oral dose, with a second dose possible 3-4 weeks later
Risks	Psychosis, mania, and risk of boundary violations in therapy
Side Effects	Headache, nausea, anxiety, dizziness
Half-Life	2-4 hours
Interactions	Psilocybin is a pro-drug that is activated by conversion into psilocin (by the enzyme alkaline phosphatase) Serotonergic medications (risk of serotonin syndrome) Lithium and tramadol (increased seizure risk)

Key Takeaways

- Psychedelics bring rapid psychological changes, making people more open, flexible, and ready to engage in psychotherapy.
- They can also cause hallucinations and a frightening sense of existential panic.
- While they are not ready for clinical use, patients are starting to experiment with them, and we can guide them to do so in a safer and more therapeutic way.

Natural Therapies and Deficiency States

CHAPTER 35

Overview of Natural Therapies and Deficiency States

THERE ARE DOZENS OF NATURAL THERAPIES for depression, but that is not what this section is about. This section focuses on a few that—in theory—address deficiencies:

- Folate and vitamin D
- Omega-3 fatty acids
- Probiotics
- Light therapy

Although most of these were born out of a deficiency model, recent research has blurred that distinction. Light therapy treats seasonal winter depression but works in the summer as well. Folate works when levels are low, as well as when they are normal.

The implication is that these deficiencies are either widespread in depression, or supraphysiologic doses have antidepressant effects. There is evidence to support both theories, but it leans toward the widespread deficiency side. Although there is plenty of light in the summer, people who stay indoors most of the time may still have a light deficiency. Even when folate levels are normal, patients may not be able to convert it to the brain-active form, l-methylfolate.

Relevance in Difficult-to-Treat Depression

These natural approaches are valuable in difficult-to-treat depression for several reasons. First, they address biological mechanisms that standard antidepressants miss. Second, they can treat depression when conventional treatments fail. Third, they carry fewer side effects and drug interactions than most psychiatric medications.

Consider these approaches when:

- Patients have not responded to one or more antidepressant trials
- Medical comorbidities complicate medication choices
- Patients are particularly sensitive to side effects
- Lab tests reveal specific deficiencies
- Patients express a preference for natural approaches

Some clinicians avoid natural therapies out of a concern that recommending them will cause patients to stop their medication. On the other hand, patients may distrust clinicians who view medications as the sole solution. Many take supplements on their own and bringing that into the treatment plan with evidence-based guidance enhances trust.

Another concern is that natural therapies are not subject to the same manufacturing standards as medications. In one study, 70% of melatonin products did not contain the amount of melatonin on the label, and many deviated widely (Erland LA and Saxena PK, *J Clin Sleep Med* 2017;13(2):275–281). To protect against that risk, I have identified products in this section that were tested in clinical trials or by independent labs.

Summary of the Evidence

Not all natural approaches are created equal. Below is a brief summary of the evidence for each therapy covered in this section:

TABLE 35-1. Summary of Natural Therapy Evidence

Therapy	Effect Size	Evidence Quality	Works as Monotherapy?	Works as Augmentation?
Light therapy	Medium to large	Strong	Yes	Yes
L-methylfolate	Small to medium	Moderate to strong	Possibly (as Enlyte)	Yes
Omega-3 fatty acids	Small to medium	Moderate	Mixed data	Yes
Probiotics	Small to medium	Preliminary but promising	Limited data	Yes
Vitamin D	Small to medium	Mixed results	Only in deficiency	Only in deficiency

These effect sizes compare favorably with conventional pharmacologic augmentation strategies. When that efficacy is combined with their favorable side effect profiles, these approaches deserve serious consideration in treatment-resistant cases.

Practical Considerations

For self-administered interventions like supplements, patient adherence is key. Here are some ways to enhance that:

- Explain the evidence and mechanism of action.
- Emphasize that the difference between supplements and medications is based more on legal categorizations than scientific ones (lithium is a medication, while magnesium is a supplement; melatonin is a prescription in most countries and valproate [Depakote] was originally extracted from the *Valeriana officinalis* plant).
- Describe stories of successful recoveries.
- Write the directions as a prescription (even if taking over-the-counter).

Set clear expectations for response. Most of these therapies require four to eight weeks for full effect, though some (like light therapy) work more rapidly.

Unlike many psychiatric medications, formal laboratory monitoring is generally not required with these treatments. However, baseline testing for specific deficiencies can help target interventions, particularly for folate, vitamin B12, and vitamin D.

Finally, there are cost considerations. Most of the options in this chapter are available over-the-counter for $10–$20 a month. Some insurers cover them, and most are available by prescription.

CHAPTER 36

Light Therapy

LIGHT THERAPY DEVELOPED out of a deficit model, and it was a personal one. When Norman Rosenthal moved from South Africa to New York City to start a psychiatric residency at Columbia, he noticed a change. In the summer, he had boundless energy, but when winter came, he felt drained, and couldn't imagine how he had done so much work in the season before. From Columbia, Dr. Rosenthal went to work at the National Institute of Mental Health, where he met a patient in 1980 who experienced severe, recurrent winter depressions. The patient improved with artificial bright light.

That launched the first controlled trial of bright light therapy for seasonal depression. It was published in 1984, and its success has been followed by nearly 100 randomized controlled trials (Mårtensson B et al, *J Affect Disord* 2015;182:1–7). But this was more of a rediscovery. Hippocrates called it heliotherapy and built solariums to improve energy and well-being. In the 1800s, Florence Nightingale noticed that patients recovered faster when their hospital rooms had light-filled windows. Her work inspired the sunrooms that briefly adorned Victorian hospitals until concerns about the spread of germs boxed everyone back in. In 1903, Niels Finsen was awarded the Nobel Prize for his work on artificial light therapy in psoriasis.

Evidence in Depression

A common misconception is that light therapy is lightweight. In seasonal depression, its effect size is medium to large (0.5–0.8), and in non-seasonal depression it is medium (0.5) (Mårtensson B et al 2015; Tao L et al, *Psychiatry Res* 2020;291:113247). When compared head-to-head with antidepressants, light therapy has similar benefits but a faster onset.

Although light therapy is associated with seasonal depression, a quarter of the controlled trials involved non-seasonal depression. In a Canadian study, light therapy had a large effect in nonseasonal depression while fluoxetine (Prozac) failed to separate from placebo (Lam RW et al, *JAMA Psychiatry* 2016;73(1):56–63).

Light therapy also treated bipolar depression in four controlled trials, generally without regard to season, with a medium effect size of 0.4 (Lam RW et al, *Can J Psychiatry* 2020;65(5):290–300).

These effects mean that the casual observer should be able to detect the difference. Those changes may be noticeable after a few days, though it takes up to a month to see the full effect of light therapy.

When to Consider Light Therapy

Light therapy is a first-line treatment for seasonal affective disorder, winter type. Around one in five people with depression have this pattern, though this varies by climate. In the US, the rate falls as we move south, particularly between the 34° and 30° northern latitudes (eg, South Carolina to Florida). Closer to the equator, there is no longer a winter pattern and instead a mild bump in summer depression. Moving further south, research is sparse, but suggests a return to the winter pattern as we approach the far end of the Southern Hemisphere in Chile and Tasmania, where the depressing drop in sunlight occurs in June and July (Nevarez-Flores AG et al, *J Psychiatr Res* 2023;162:170–179).

In seasonal depression, atypical features like hypersomnia and high appetite predict response to light (Terman M et al, *Am J Psychiatry* 1996;153(11):1423–1429). For nonseasonal depression, it's not as clear who will respond, but light therapy is worth considering when patients cannot tolerate or do not respond to antidepressants. Light therapy is effective in populations where antidepressants are not ideal, including:

- Pregnant women
- Adolescents
- Traumatic brain injury
- Bipolar depression

Light therapy works as both monotherapy and augmentation in nonseasonal depression. It also augments the antidepressant effects of transcranial magnetic stimulation (TMS) (Barbini B et al, *Int J Psychiatry Clin Pract* 2021;25(4):375–377).

A few comorbidities respond to light therapy. When used in the winter months, light therapy improves bulimia nervosa and low libido in men (testosterone declines in winter) (Braun DL et al, *Compr Psychiatry* 1999;40(6):442–448; Bossini L et al, *Psychother Psychosom* 2009;78(2):127–128). When used without regard to season, light therapy improves primary insomnia, combat-related post-traumatic stress disorder (PTSD), negative symptoms of schizophrenia, fibromyalgia, and back pain (Youngstedt SD et al, *Mil Med* 2022;187(3-4):e435-e444).

The protocols for treating these comorbidities are similar to that for depression. To treat insomnia though, light is delivered very early in the morning (4:00–5:00 am). Light therapy can also improve sleep and energy in shift-workers, where it is delivered shortly after waking up.

Mechanism of Action

Light therapy synchronizes the abnormal circadian rhythms that are part of the pathophysiology of mood disorders. It alters serotoninergic and noradrenergic transmission; regulates cortisol and melatonin; enhances parasympathetic activity; and reduces amygdala hypersensitivity.

Evidence for a Deficit

There is a clear sunlight deficit during the winter months, but how does light therapy work in the summer? Indoor living is a possibility. Even the brightest indoor spaces only deliver about 500 lux of light, compared to 1,000 lux for an overcast day and 10,000–30,000 lux for a sunny one. Effective light boxes are generally in the 10,000 lux range.

How to Use Light Therapy

Selecting a Light Box

Most light boxes do not work. The qualities that make them attractive—slim, portable, and unobtrusive—also prevent them from giving off enough light to treat depression. An effective box needs a large screen (at least 12 × 17 inches) that hangs above the head and gives off intense white-spectrum light.

Just how bright should the light be? Ideally 10,000 lux. Lower doses can also work, but they require more time (eg, 2–4 hours for a 2,500 lux box vs 30–60 minutes for 10,000 lux). The box also needs a filter to screen out dangerous ultraviolet rays. I keep updated recommendations at www.

chrisaikenmd.com/lighttherapy. The Center for Environmental Therapeutics is another good resource (www.cet.org).

There is not as much evidence to support alternatives to the traditional light box, including light visors, light glasses, full spectrum boxes, and blue-light lamps.

Timing the Treatment

Light therapy works in part by setting the biological clock, and morning light has the most potent effect on that circadian system. The sweet spot for light therapy is generally between 5:00 am and 8:00 am, and depends on whether the patient is a morning person (closer to 5:00 am is ideal) or a night owl (closer to 8:00 am is ideal). The Center for Environmental Therapeutics has developed a self-report scale to predict the optimal start time (the Morningness-Eveningness Questionnaire or AutoMEQ at www.cet.org, under the Assessments tab).

Most patients with depression need 30–60 minutes a day of light therapy at 10,000 lux (or 2–4 hours at 2,500 lux). They should start to see improvement within one to two weeks. If they have not recovered after four weeks, try increasing the duration. Patients with a strong seasonal pattern should start light therapy preventatively two weeks before the typical start of their winter episodes.

Positioning the Box

The goal in all this is to imitate the sun. That means sitting with the light tilted at a 30° angle above the head. Patients can read, eat, use a computer, or meditate under the light, but should avoid looking directly into it for the same reasons they shouldn't stare at the sun. They can wear glasses as long as they don't have transition, blue-blocking, or tinted lenses. The intensity of the light falls exponentially with distance, so their head should stay 10–14 inches from the screen, depending on the model.

FIGURE 36-1. Positioning the Lightbox

TABLE 36-1. Light Therapy Summary

FDA Approval	None
Off-Label Benefits	Seasonal and nonseasonal depression In winter: bulimia, low libido in men In any season: bipolar depression, PTSD, insomnia, shift-work disorder
Dose	30-120 min in morning (5:00 am-8:00 am)
Risks	Mania Photosensitivity; ocular risks in patients with glaucoma, cataracts, retinal detachment, or retinopathy
Side Effects	Headache, eye strain, insomnia (if used too late)
Interactions	Caution with meds that cause photosensitivity (lamotrigine, tricyclics, antipsychotics)
Recommended Products	Carex Daylight Sky or Elite with LED bulbs (see chrisaikenmd.com/lighttherapy)
Cost	$100-$220 for box

Troubleshooting

Many patients are skeptical of light therapy. I'll emphasize that it's as effective as an antidepressant and alters neurotransmitters like serotonin, dopamine, and melatonin. Others believe that light is beneficial but think a sunroom or bright reading lamp will suffice. All of those are helpful, but the intensity of light they emit is similar to the sham boxes used as a placebo in the light therapy trials (300–1,000 lux). On the other hand, morning aerobics or a one-hour outdoor walk in the winter may work in seasonal depression.

Timing is another obstacle. If early morning is not practical, patients can still benefit by using a light box later in the day, as long as it's before 2:00 pm. After that time, light is not going to set the biological clock and may flip it in the wrong direction, causing depression, mania, and insomnia. Patients who have difficulty getting out of bed to start light therapy can benefit from a dawn simulator, which improves morning wakefulness and has a small benefit in seasonal, winter depression.

Side Effects

Tolerability

Light therapy is well tolerated. Headaches, eyestrain, and mild nausea are the most common adverse effects.

Risks

Exposure to high intensity light can damage the skin and eyes. Reports of actual problems are very rare, but patients should consult with their ophthalmologist if this is a concern (Brouwer A et al, *Acta Psychiatr Scand* 2017;136:534–548). Recommended boxes have a diffusion screen that filters out ultraviolet light, the most harmful ray. Blue light, which lies next to the ultraviolet spectrum, will still pass through and may pose a problem for patients who have retinal disease or take photosensitizing medications like lamotrigine (Lamictal), antipsychotics, or tricyclics.

Key Takeaways

- Light therapy treats depression in the winter and summer months. Its benefits are similar to those of antidepressants but its onset is faster.
- Most light boxes on the market do not have the right specifications.
- Morning light (5:00 am–8:00 am) works best in depression; avoid using the light box after 2:00 pm.
- Light therapy is well tolerated and improves some comorbid disorders (PTSD, bulimia, insomnia, and low libido in men).

CHAPTER 37

Folate, L-Methylfolate, and SAMe

THE FOLATE STORY BEGINS IN 1931. Working in tropical India, the British hematologist Lucy Wills discovered that a yeast extract, Marmite, treated macrocytic anemia. She later identified folate as the key ingredient. Her work was recognized with a full diploma from her alma mater, Cambridge University, which had previously only granted certificates to its female graduates.

In the 1960s, new technology made it easy to measure folate in clinical practice. Physicians soon discovered a host of illnesses linked to folate deficiency, including neuropathy, neural tube deficits, and dementia. Psychiatrists found low folate in a quarter of hospitalized patients, particularly in those with psychosis and depression. Folate deficiency proved to be a strong predictor of antidepressant nonresponse and supplementation an effective treatment for these cases.

In later trials, folate proved effective in general depression, regardless of baseline folate levels. Several compounds in the folate cycle have been tried—folic acid, folinic acid, l-methylfolate, and s-adenosylmethionine (SAMe)—and one has come out the clear winner: l-methylfolate.

Evidence in Depression

Evidence from nine controlled trials enrolling 6,707 patients supports the use of l-methylfolate in depression. From those studies, we know that it augments antidepressants with an effect size of 0.4, similar to that of other pharmacologic augmentation strategies (Maruf AA et al, *Pharmacopsychiatry* 2022;55(3):139–147). The 15 mg dose is more effective than 7.5 mg. All of these trials tested it as augmentation, but it did have powerful effects as monotherapy when used as part of a B-complex vitamin (Enlyte) in patients with genetic deficiencies of folate metabolism, at the methylenetetrahydrofolate reductase (MTHFR) gene.

Enlyte

Enlyte is a branded vitamin complex cleared by the FDA for "dietary management of major depressive disorder." The main ingredient is l-methylfolate (7 mg), but it also includes folic acid, folinic acid, B1, B2, B3, B6, and B12, as well as other nutrients with potential benefits in depression (iron, magnesium, zinc, coenzyme Q10, and omega–3 fatty acids). Although the doses included are low, Enlyte delivers them in more potent, bioactive forms.

In a large, industry-sponsored trial of patients with depression and MTHFR gene abnormalities, two months of Enlyte achieved a large effect size of 0.9 compared to placebo. Unlike most vitamin trials, Enlyte was tested on its own, without an antidepressant, in a mix of treatment-naïve and antidepressant-resistant patients (Mech AW and Farah A, *J Clin Psychiatry* 2016;77(5):668–671). The impressive effect size shows how powerful l-methylfolate can be in the presence of MTHFR abnormalities (in this case, the C677T or A1298C polymorphism).

Enlyte's manufacturer, JayMac Pharmaceuticals, also makes EnBrace HR. The two products are identical except that EnBrace HR is marketed as a prenatal vitamin for women at risk for depression. It is a promising direction, but so far, there has only been a small open-label EnBrace HR trial in this population.

Folic acid, Folinic acid, and SAMe

What about other forms of folate? Folic acid had promising but mixed results in small trials, but then it failed in a large trial of 475 patients (Sarris J et al, *Am J Psychiatry* 2016;173(6):573–587). Folinic acid was only tested in an uncontrolled trial, and even there it didn't clearly work.

SAMe is not a folate vitamin, but it plays an important role in the folate cycle. L-methylfolate is converted into SAMe, which is then used to produce neurotransmitters like serotonin, norepinephrine, and dopamine. SAMe looked promising in early trials that compared it to tricyclics and selective serotonin reuptake inhibitors (SSRIs), but those trials are prone to false results because they did not include a placebo group. When compared to a placebo, SAMe has largely failed to treat depression (Peng TR et al, *Gen Hosp Psychiatry* 2024;86:118–126).

One explanation for the negative findings may be in the product itself. SAMe is more temperamental than most supplements, requiring blister packaging to protect it from heat and moisture, and enteric coating to prevent stomach acids from degrading it.

When to Consider L-Methylfolate

With its low cost, solid evidence, and favorable tolerability, l-methylfolate belongs at the top of the list of therapies for difficult-to-treat depression. Patient factors that predict a positive response to l-methylfolate are listed in the table and include obesity, inflammation, and polymorphisms of the MTHFR gene.

TABLE 37-1. Predictors of L-Methylfolate Response

MTHFR variation (homozygous or C677T)
Obesity (BMI ≥ 30 kg/m²)
Inflammation (hs-CRP ≥ 3 mg/L)
Pregnancy and lactation
Eating disorders
Renal failure and GI disease
Poor nutrition
Alcohol use disorders
Smoking
Concomitant medications (eg, lamotrigine, valproate, carbamazepine, phenytoin, fluoxetine, oral contraceptives, methotrexate, metformin, sulfasalazine, warfarin, and triamterene)

Mechanism of Action

Folate is one of the building blocks for the neurotransmitters that regulate mood: serotonin, norepinephrine, and dopamine. L-methylfolate is the active form in the brain. Folic acid and folinic acid are both precursors of l-methylfolate. They are similar, except that folic acid is synthetic and folinic acid is natural. They are converted into l-methylfolate by the MTHFR enzyme. Some people have a genetic deficiency in MTHFR that slows this conversion, which explains why they may require the l-methylfolate form.

Another way that folate may treat depression is by lowering homocysteine. When folate levels are low, homocysteine levels rise, and elevated homocysteine is linked to heart disease, stroke, osteoporosis, depression, and cognitive decline.

TABLE 37-2. L-Methylfolate Summary

FDA Approval	Cleared as a dietary supplement for depression
Off-Label Benefits	Bipolar depression, autism, negative symptoms of schizophrenia
Dosing	15 mg QD
Risks	None
Side Effects	None
Interactions	None
Recommended Products	Prescription form or over-the-counter as MethylPro or Opti-Folate; also available as a part of a proprietary multivitamin (Enlyte, EnBrace HR)
Cost	\$8-\$30/month

Evidence for a Folate Deficit

Folate deficiency is linked to depression, as are deficiencies of the enzymes that activate folate. MTHFR mutations moderately increase the risk of depression (odds ratio 1.4) (Gilbody S et al, *Am J Epidemiol* 2007; 165(1):1–13). In the general population, the prevalence of MTHFR mutations is about 30%–40%, but patients with depression have a higher prevalence—around 60% (Mischoulon D et al, *CNS Spectr* 2012;17(2):76–86). These mutations are also more common in people of Hispanic or Mediterranean descent.

However, MTHFR mutations are not the only reason that patients respond to l-methylfolate. Low folate levels are common in depression (10%–20%), as are factors that impair folate metabolism, like inflammation, obesity, smoking, and alcohol use. Adding all these together suggests that most patients with depression have a reason to consider l-methylfolate.

How to Use L-Methylfolate

L-methylfolate is the clear winner in the folate category, but should you recommend it on its own or as a compounded multivitamin (Enlyte)? Both options have sound theory and good clinical data behind them, but we lack comparative studies to clarify the choice. One strategy is to have patients compare mood ratings after a one- to two-month trial of each version. Enlyte's manufacturer offers a coupon that helps defray the cost at www.enlyterx.com.

Both versions are well tolerated and can be started at the treating dose (l-methylfolate 15 mg QD or Enlyte 1 capsule QD). Lower doses of l-methylfolate (7.5 mg) also work, but are not as effective as the 15 mg. Unlike Enlyte, l-methylfolate is available as a generic prescription (previously branded Deplin) or over-the-counter.

Side Effects and Risks

L-methylfolate is not associated with serious risks or tolerability problems. Like other folate supplements, it may mask symptoms of vitamin B12 deficiency.

Key Takeaways

- Folate is used in the synthesis of serotonin, norepinephrine, and dopamine.
- Impairments of folate metabolism are common in depression, and the causes include low serum levels as well as genetic deficits, inflammation, and drug interactions.
- L-methylfolate, and the related vitamin complex Enlyte, are FDA cleared as nutritional supplements for depression. Evidence from large randomized controlled trials supports their use.
- Folic acid, folinic acid, and SAMe are also involved in the folate pathway, but their evidence in depression is less robust than for l-methylfolate.

CHAPTER 38

Vitamin D

VITAMIN D IS NOT A VITAMIN. It is a neurosteroid that helps us absorb calcium, which is necessary for bone strength. We can maintain our stores of vitamin D in two ways: by taking supplements or exposing ourselves to bright, midday sunlight for at least 15 minutes three times per week. Long-term vitamin D deficiency causes a range of health problems, including bone fractures, rickets, and possibly cancer, cardiovascular disease, and earlier death (Theodoratou E et al, *BMJ* 2014;348:g2035). More controversial is whether less extreme deficiencies (insufficiencies) of vitamin D can cause more subtle problems with energy or mood.

Evidence in Depression

Unlike the other treatments in this section, vitamin D only treats depression in patients with low serum levels. In patients with a true deficiency (< 20 ng/mL), vitamin D supplementation improves mood with a small to moderate effect size. On the other hand, vitamin D supplementation is not helpful for depressed patients with normal vitamin D (Menon V et al, *Indian J Psychol Med* 2020;42(1):11–21).

When to Consider Vitamin D

Consider vitamin D when levels are below 20 ng/mL, and check levels in patients with treatment resistance. Besides depression, risk factors for low vitamin D include older age, obesity, lack of sunlight, malabsorption, and dietary deficiencies.

Mechanism of Action

Although its precise mechanism is unknown, there are several ways that vitamin D might influence depression. It regulates stress hormones and

TABLE 38-1. Vitamin D Summary

FDA Approval	Vitamin D deficiency, corticosteroid-induced osteoporosis (with calcium), refractory rickets, familial hypophosphatemia, hypoparathyroidism (with calcium)
Off-Label Benefits	Depression in vitamin D deficiency
Dosing	Loading dose based on deficiency level, then 800-2,000 IU daily maintenance
Risks	High levels (> 60 ng/mL) may cause hypercalcemia and kidney stones
Side Effects	Minimal at standard doses
Interactions	May increase absorption of calcium and aluminum (relevant for patients on antacids)
Recommended Products	Vitamin D3 preferred over D2; available prescription or over-the-counter
Cost	$3-$10/month

plays a role in brain growth and neuronal development (Eyles DW et al, *Front Neuroendocrinol* 2013;34(1):47–64).

Evidence for a Vitamin D Deficit

Vitamin D deficiency (< 20 ng/mL) is remarkably common, affecting around 1 in 4 US adults. In depression, the rate is higher: 1 in 3 adults (Boerman R et al, *J Clin Psychopharmacol* 2016;36(6):588–592). In epidemiological studies, low levels of vitamin D are associated with autism, schizophrenia, seasonal affective disorder, and depression.

How to Use Vitamin D

To check vitamin D status, order a total 25-Hydroxy Vitamin D level (billing code E55.9). Treatment depends on the severity of deficiency:

- For levels below 10 ng/mL (severe deficiency): Vitamin D3 50,000 IU weekly for 2–3 months
- For levels between 11–25 ng/mL (moderate deficiency): Vitamin D3 10,000–20,000 IU weekly for two to three months
- After the loading phase, recheck levels at 3 months, aiming for 30–40 ng/mL
- For maintenance, continue with 800–2,000 IU daily and recheck levels every 6 to 12 months

- Keep serum level below 50–60 ng/mL, as higher levels increase the risk of levels above 60 ng/mL may increase risk of hypercalcemia, kidney stones, and other adverse effects.

Side Effects and Risks

Vitamin D is well tolerated. While supplementation improves health when levels are low, there may be risks with going too high (eg, > 50 ng/mL). Those risks are unproven, but showed up in observational studies where high levels of vitamin D were linked to respiratory infections, cancer, and earlier death. In terms of mortality, the sweet spot appears to be between 20 and 40 ng/mL (Durup D et al, *J Clin Endocrinol Metab* 2012;97(8):2644–2652).

Key Takeaways

- Vitamin D supplementation can improve mood when levels are low (< 20 ng/mL) but otherwise does not have a clear benefit.
- Besides depression, low vitamin D is linked to poor bone health, cancer, heart disease, and early death, but raising the levels too high may be harmful to health.

CHAPTER 39

Omega–3 Fatty Acids

OMEGA–3 FATTY ACIDS ("FISH OIL") improve mood and decrease irritability in a wide range of populations, including autism, borderline personality disorder, and substance use disorders, as well as people without mental illness (eg, prisoners, married couples, teens, and elderly Thai men) (Sinn N et al, *Nutrients* 2010;2(2):128–170). Its origins date back to the 1700s, when English physicians began using cod liver oil for various ailments, though some find traces in the writings of Hippocrates and the Bible, where an angel in the book of Tobias describes fish liver as a "useful medicine."

The modern era of omega–3s began in the 1970s with two lines of research that clarified the importance of these healthy fats in the brain and heart. In the neurosciences, researchers discovered that animals raised without omega–3s developed nerve cells demyelination and profound brain impairments. In clinical and epidemiologic studies, low levels of omega–3s correlated with more severe depression.

Meanwhile, cardiologists became intrigued by a puzzling phenomenon in Greenland. Heart disease was exceedingly rare among the Inuits who lived there, despite their high-fat diet. Danish physician Jørn Dyerberg linked this to the heart-healthy omega–3s that were plentiful in the seafood they consumed. These fats improved lipid balance, leading the FDA to clear them as a food supplement for lowering triglycerides in 2004.

Evidence in Depression

The first clinical trial in psychiatry appeared in 1999, where omega–3s prevented new episodes in bipolar disorder. Studies in unipolar depression soon followed, and there are now more than 40 randomized controlled trials involving around 10,000 participants. Some of those studies were negative, but there are patterns in the results that inform practice (Hallahan B et al, *Br J Psychiatry* 2016;209(3):192–201).

- Omega–3s have clear antidepressant effects in trials that enrolled patients with DSM-based major depressive disorder but are less clear when used for nonspecific depressive symptoms.
- Omega–3s treat depression in trials that use a specific ratio of fatty acids.

There are two types of omega–3s: EPA and DHA (eicosapentaenoic acid and docosahexaenoic acid). The formulations that work are either pure EPA or contain at least twice as much EPA as DHA (Guu TW et al, *Psychother Psychosom* 2019;88(5):263–273). These formulations effectively augment antidepressants with a medium effect size (0.5–0.6), similar to the effect of the average psychiatric treatment.

One controlled trial points to patients with inflammation as better responders (ie, those with a high-sensitivity C-reactive protein level ≥ 3 mg/L). In that study, only the depressed patients with elevated inflammatory biomarkers responded to omega–3s, and the higher the inflammation, the better the response (Rapaport MH et al, *Mol Psychiatry* 2016;21(1): 71–79). In a separate study, patients with inflammation responded better to high dose (EPA 4,720 mg with DHA 1,160 mg) than lower doses (EPA 1,180–2,360 mg with DHA 254–508 mg) (Mischoulon D et al, *J Clin Psychiatry* 2022;83(5):21m14074).

When to Consider Omega–3 Fatty Acids

After depression, the best evidence for omega–3s is in ADHD, borderline personality disorder, bipolar disorder, and schizophrenia. If we look for patterns in these various disorders, omega–3s improve emotional lability and decrease irritability and impulsivity, but they also reduce core symptoms of ADHD and prevent psychosis in youth at risk for schizophrenia.

Omega–3s are also useful for common physical health problems. Several prescription omega–3s are FDA-indicated for lowering triglycerides. They also have evidence in hypertension, inflammatory bowel disease, dry eyes, macular degeneration, fatty liver, arthritis, asthma, and psoriasis. The doses in medical disorders are often on the high side (3,000–4,000 mg/day of EPA+DHA for psoriasis and hypertriglyceridemia).

Omega–3s improve lipid balance in people on antipsychotics, and they may be useful for two side effects of lithium: psoriasis and acne.

Mechanism of Action

Omega–3s are a rare example where the mechanism makes intuitive sense in a way that engages patients. I explain it like this, "Omega–3s comprise 30% of the human brain, where they coat and protect brain cells. The brain is derived from the same embryonic cells as the skin, and like the skin—or like a leather chair—it needs oil to stay supple and healthy. Most people don't get enough omega–3s from their diet, forcing the brain to substitute other oils like cholesterol. On neuroimaging, the brain cells are less flexible, and the person is less flexible as well—more irritable and depressed."

That neuronal flexibility is based on a study of T(2) whole brain relaxation times in patients treated with omega–3s (Hirashima F et al, *Am J Psychiatry* 2004;161(10):1922–1924). Omega–3s also enhance serotoninergic and dopaminergic transmission, reduce inflammation, promote neuroprotection, and modulate glucocorticoids in the stress response.

Evidence for a Deficit

Omega–3s are essential fatty acids, which means they can only be obtained from the diet. Dietary intake has declined in the past 150 years as people have shifted away from the oily fish, nuts, and vegetables that supply them. Besides diet, genetic differences in fatty acid metabolism can also cause deficits. At both a population and individual level, lower omega–3s predict more depression, but psychiatric trials have not examined whether changes in omega–3 levels correlate with treatment response (Grosso G et al, *Oxid Med Cell Longev* 2014;2014:313570).

How to Use Omega–3 Fatty Acids

Omega–3s are well tolerated and can be started at the treating dose. The standard dose for depression is 1,000–4,000 mg daily of EPA + DHA, using a product with an EPA:DHA ratio of at least 2:1. Higher doses (toward 4,000–5,000 mg daily of EPA + DHA, with at least 4,000 mg EPA) are recommended for patients with inflammation.

Gastrointestinal side effects can be improved by refrigerating the capsules or using an enteric-coated or odor-neutralized product. A high-fat meal increases their absorption by approximately 25%.

Omega–3s are available by prescription or over-the-counter, and both routes have problems. Most over-the-counter preparations are low in EPA. In the table are high-EPA products that passed independent tests for integrity (through Consumer Labs or US Pharmacopeia). Only one prescription product contains enough EPA to match the necessary ratio (Vascepa). However, prescription omega–3s are in an estherized form, which lowers their bioavailability by 30% and may increase their side effects (Dyerberg J et al, *Prostaglandins Leukot Essent Fatty Acids* 2010; 83(3):137–141).

Unfortunately, I have not found vegetarian options with enough EPA for psychiatric use. Some products also contain omega–7s, which are found in the diet and do not have significant risks or benefits.

Dietary Sources

Technically, patients could get a clinical dose of omega–3 in their diet by eating two to three servings a week of oily fish, like salmon, herring, or anchovies (9–12 ounces per week). Other dietary sources include walnuts, flaxseeds, edamame, fruits, and dark green vegetables.

Side Effects

Tolerability

Omega–3s are well tolerated with side effects of nausea, belching, and fishy taste.

Risks

Omega–3s may prolong bleeding time, a problem for patients undergoing surgery or taking anticoagulants, but this risk was not detectable at the doses used in psychiatry (< 4,000 mg/day) (Mori TA, *Food Funct* 2014; 5(9):2004–2019). To be on the safe side, stop them two weeks before surgery, and restart after the procedure (they improve wound healing).

In high doses, omega–3s may cause arrhythmias in patients with atrial fibrillation. One study found a correlation between omega–3 serum levels and prostate cancer, but other research refutes this or even shows the opposite effect, and the association might be explained by contaminants in dietary fish (Aucoin M et al, *Integr Cancer Ther* 2017;16(1):32–62). Contamination by heavy metals like mercury is a concern with fish oil, and the products recommended in the table were tested for metal contamination.

TABLE 39-1. Omega–3 Fatty Acids Summary

FDA Approval	Elevated triglycerides
Off-Label Benefits	Major depression, bipolar depression, borderline personality disorder, ADHD, autism, negative symptoms of schizophrenia, irritability Psoriasis, acne, fatty liver
Dosing	1,000-4,000 mg QD of EPA + DHA Higher doses for inflammation and psoriasis Use a product with an EPA:DHA ratio ≥ 2:1
Risks	Anticoagulant effects (stop 1 week prior to surgery, restart after surgery); avoid in CHF and atrial fibrillation
Side Effects	Reflux, fishy taste, diarrhea
Interactions	None
Cost	Prescription Vascepa is $30/mth. OTC versions are $5-20/mth.

TABLE 39-2. Recommended Omega–3 Products

Product	EPA/DHA in a capsule	Monthly cost for starting dose (1,000 mg/day)
High concentration (starting dose = 1 capsule)		
Carlson Elite EPA Gems	1,000/16.5 mg	$15
GNC Triple Strength Enteric Coated	734/266 mg	$8
Nature Made 1400	683/252 mg	$9
Nordic Naturals ProOmega (Liquid)	1,950/975 mg	$22
Nutrigold Triple Strength	725/275 mg	$11
Omegavia Ultra Concentrated Enteric Coated	780/260 mg	$15
Spring Valley Max Care 2000 (Walmart)	645/310 mg	$4
Vascepa (Rx only)	960/40 mg	$30
Viva Naturals Triple Strength	750/142.5 mg	(monthly cost) $9
Low concentration (starting dose = 2 capsules)		
OmegaVia EPA 500	500/21.2 mg	$15
Vitacost Super EPA (Vitacost.com)	600/150 mg	$16

Key Takeaways

- The Western diet is deficient in omega–3 fatty acids, which play an important role in the formation of neuronal cell membranes.
- Supplementation with omega–3s treats depression and improves metabolic health, but most commercial products do not have the right EPA:DHA ratio (≥ 2:1).
- Omega–3s improve emotional lability and decrease irritability and impulsivity across multiple psychiatric disorders.

CHAPTER 40

Probiotics

A WALK DOWN THE GROCERY AISLES gives the impression that probiotics are a fad. Their health benefits are touted in yogurt, kombucha, fermented foods, and sourdough breads. In fairness, this fad has deep roots. When Louis Pasteur developed the germ theory of illness in 1861, he knew that microbes were plentiful in the human body and believed they could do both good and harm. In the early 1900s, Nobel laureate Elie Metchnikoff observed that people lived longer when they ate yogurt regularly and linked this to the probiotic *Lactobacillus*. Clinical trials soon followed, and by the 1920s probiotics were known to improve constipation, diarrhea, eczema, and mental health.

With the discovery of penicillin in 1928, research shifted from probiotics to antibiotics. Probiotics gained attention again in the early 2000s, when researchers proved that an unhealthy gut flora was not just the result of illness but could directly cause it. They did this by transplanting the microbiome from an obese mouse into a healthy mouse. After the fecal transplant from an obese mouse, the healthy mouse gained significant weight, while transplants from a healthy mouse had no such effect. This experiment was repeated with ulcerative colitis, colorectal cancer, anxiety, and depression (Kelly JR et al, *J Psychiatr Res* 2016;82:109–118).

Those studies would not be ethical in humans, but human studies have found the reverse. Mood improves when the microbiome of a healthy person is transplanted into a patient with depression. We see the same effect with disorders of anxiety, insomnia, autism, and alcohol use (Vasiliu Ol, *Curr Med Res Opin* 2023;39(1):161–177).

Fecal microbiota transplantation is experimental in psychiatry,* but probiotic supplementation operates on the same principle. It is becoming

* In 2022, the FDA approved the first fetal microbiota transplant for *C. difficile* infection. It is administered as a capsule.

mainstream with the appearance of well-designed controlled trials and meta-analyses which we'll review next (Nikolova VL et al, *JAMA Psychiatry* 2023;80(8):842–847; Merkouris E et al, *Microorganisms* 2024;12(2):411).

Evidence in Depression

Probiotics are the good bacteria in the gut microbiome. Starting in the early 2010s, trials of probiotics found benefits in depression, anxiety, cognition, and sleep. The early trials focused on people who had mild depression related to stress or medical problems. Trials in major depressive disorder came next, and probiotics proved more effective here than in studies of stressed-out subjects.

There are at least a dozen randomized controlled trials of probiotics in clinical depression, including moderate and inpatient depression, of which 80% are positive. What is missing from this data is size. They average 67 subjects per trial, and there are no large trials among them. Besides depression, probiotics improved anxiety and cognitive functioning in these trials. They also worked in treatment-resistant cases, where they augmented antidepressants after failure of one or two antidepressant trials (Miyaoka T et al, *Clin Neuropharmacol* 2018;41(5):151–155).

Most antidepressants worsen microbiome diversity, so combining them with a probiotic makes sense (Borgiani G et al, *Int Clin Psychopharmacol* 2025;40(1):3-26). Antipsychotics also disturb the microbiome, which contributes to their metabolic side effects. Weight gain, cholesterol, and insulin resistance are lessened when patients take probiotics with antipsychotics, particularly if they take them with prebiotic fiber (Huang J et al, *Transl Psychiatry* 2022;12(1):185).

When to Consider Probiotics

We don't know who is more likely to respond to probiotics, but a good place to start is those with signs of poor microbiome health. That includes depression itself, but also:

- Western-style diet
- Sedentary lifestyle
- Insomnia
- Antipsychotic or antibiotic use
- Inflammation and obesity

Patients with medical disorders that respond to probiotics are also good candidates. These include diabetes type I and II, vascular disease, irritable bowel syndrome, Crohn's disease, ulcerative colitis, multiple sclerosis, dermatitis, and eczema.

Mechanism of Action

Probiotics influence the brain through the two-way communication of the gut-brain axis. They act as tiny neurotransmitter factories, producing serotonin, dopamine, norepinephrine, glutamate, gamma-aminobutyric acid (GABA), histamine, and acetylcholine, either directly or through their precursors. They also:

- Raise brain-derived neurotrophic factors (BDNF)
- Influence vagal nerve signaling
- Reduce inflammation
- Balance the hypothalamic-pituitary-adrenal (HPA) axis (ie, stress hormones like cortisol)

Evidence for a Deficit

Most mental illnesses, including depression, schizophrenia, autism, and addictions, are associated with poor gut health, also known as dysbiosis. The deficit here is in the variety of the flora and the number of healthy bacterial strains. Improvements in the gut flora correlate with their antidepressant effects of probiotics.

How to Use Probiotics

Polypharmacy is frowned upon in psychiatry, but with gut health, diversity is the goal. Studies show that using multiple probiotic strains is more effective than using single strains (Ferrari S et al, *J Tradit Complement Med* 2024;14(3):237–255). The strains with the best evidence in depression are those from the *Lactobacillus* and *Bifidobacterium* families*, followed by *Streptococcus thermophiles*. *Lactobacillus* is the first strain that colonizes the body through vaginal birth, and was the first to gain recognition for its health benefits.

---

*There are many subspecies within these families. The most used are *Lactobacillus*: *acidophilus, paracasei, casei, plantarum, salivarius*; to a lesser extent *helveticus, lactis, delbrueckii, rhamnosus*. *Bifidobacterium*: *bifidum, lactis, breve, longum*; to a lesser extent *infantis, subtilis*.

TABLE 40-1. Probiotics Summary

FDA Approval	None
Off-Label Uses	Major depression, bipolar disorder, anxiety, autism, cognition, dementia, negative symptoms of schizophrenia Constipation, irritable bowel syndrome, inflammatory bowel disease, dyslipidemia, weight loss, antibiotic-associated diarrhea, *C. difficile* infection
Dosing	2-10 billion CFUs
Risks	Avoid in immunocompromised
Side Effects	Gas
Interactions	None
Recommended Products	Lifted Naturals Mood Boosting Trunature Advanced Digestive
Cost	$10-$15/month

Probiotics are dosed by the number of bacteria in the capsule. This is measured in the billions, which is small potatoes considering the 100 trillion organisms that already live in the gut. Doses range from 2–10 billion colony forming units (CFUs), averaging 7 billion CFUs (ie, 7×10^9). Some are available by prescription (eg, VSL#3, Visbiome, and Restora), but these are best avoided as their CFU count is 10–100 times higher than the amounts used in psychiatry. Probiotics are packaged in capsules, alive but dormant in a freeze-dried state. Once ingested, they grow in the gut, but these are best avoided as they need the right "prebiotic" diet to flourish.

Prebiotics are fibrous nutrients that promote the growth of healthy bacteria. They are found in fruits, vegetables, beans, nuts, seeds, and whole grains. They can also be taken as prebiotic supplements (eg, inulin and galactooligosaccharide), or through a symbiotic, which is a combination of a pre- and probiotic.

Although common sense suggests probiotics work better with prebiotics, and studies outside of depression support that notion, we don't have evidence that prebiotics are necessary to treat depression. The recommended products in the table contain a prebiotic along with most of the probiotic strains that improve depression (and were tested by independent labs for integrity).

Side Effects

Tolerability

Probiotics are generally well tolerated. They can cause stomach upset, gas, bloating, and diarrhea (but they also treat antibiotic-associated diarrhea).

Risks

Probiotics were safe and well tolerated in psychiatric trials. Avoid them in patients who are severely immunocompromised, such as from HIV or chemotherapy.

Key Takeaways

- Poor microbiome health (dysbiosis) plays a causative role in depression across multiple pathways, including inflammation, vagal nerve signaling, neuroprotection, and neurotransmitter production.
- Probiotics treat depression and augment antidepressants with a small effect size.
- They have additional benefits for anxiety, cognition, and metabolic health.

Neuromodulation

CHAPTER 41

Overview of Neuromodulation

IMAGINE YOU ARE LIVING IN EGYPT around 2700 BC. Construction is beginning on the pyramids, but you've been forced to take medical leave due to a debilitating neuropathy. Walking along the Nile, you are jolted by a shock as you step on an electric eel. It's painful at first, but soon afterwards the chronic pain you've suffered goes away.

That is probably how humankind discovered neuromodulation, the use of electrical stimulation to treat neuropsychiatric disorders. Ancient texts describe similar chance encounters with eels, and Egyptian physicians recommended electric eel stimulation (EES) for similar indications as we use transcranial magnetic stimulation (TMS) for today: migraines, pain, epilepsy, and melancholy. The Greeks and Romans did much the same.

The modern era of neuromodulation began with electroconvulsive therapy (ECT). In the new millennium, the field has grown to include:

- Transcranial magnetic stimulation (TMS)
- Transcranial direct current stimulation (tDCS)
- Vagus nerve stimulation (VNS)
- Deep brain stimulation (DBS)

In this section, we'll focus on the two options that are most relevant to clinical practice: ECT and TMS. Both are FDA cleared for treatment-resistant depression. In contrast, tDCS is a milder form of electrical stimulation that has become popular as a DIY at-home treatment. There is good evidence that it treats depression, including head-to-head trials with antidepressants, but tDCS is not cleared by the FDA and lacks data in treatment-resistant depression.

VNS and DBS are effective in treatment-resistant depression, but these procedures are more invasive, requiring implantation of an electrical device in the neck and chest (VNS) or brain (DBS). VNS was cleared in 2005 for highly resistant depression—after failure of four or more antidepressant treatments. DBS has FDA clearance for Parkinson's disease, essential tremor, and epilepsy, but is investigational for depression. DBS is only considered after failure of four antidepressants and ECT, but in 2018 researchers at Stanford University brought 70% of patients with similar levels of treatment resistance to remission with a new version of TMS that we'll introduce in the next chapter (Williams NR et al, *Brain* 2018;141(3):e18).

TABLE 41-1. Neuromodulation and Interventional Psychiatry: What to Expect

Treatment	Frequency	Activity Limitations	Maintenance Schedule (optional)
ECT	1-hour sessions 3 days a week for 3-4 weeks	Cannot drive on day of session, unable to work during treatment	Monthly
TMS	3-45 min sessions* 5 days a week for 6 weeks	None	1-8 per month
fMRI-guided SNT (SAINT) TMS	10 hours a day for 5 days (with a 10-min treatment every hour)	None beyond time commitment	As needed if symptoms arise (typically 1 day per month)
Ketamine (IV) and Esketamine (IN)	30-45 min sessions Twice weekly for 4 weeks, followed by 4 weekly treatments	Wait 2-3 hours before driving	1-2 per month

*3 min for intermittent theta-burst stimulation (iTBS), 20 min for Deep TMS (H coil, Brainsway), 45 min for traditional TMS (figure-8 coil)

CHAPTER 42

Transcranial Magnetic Stimulation (TMS)

TRANSCRANIAL MAGNETIC STIMULATION (TMS) has had trouble finding its footing, but there are reasons to consider it more often in depression. Compared to pharmacologic augmentation, TMS is more effective and better tolerated. Unlike electroconvulsive therapy (ECT), TMS does not worsen cognition and may improve it.

TMS has undergone significant changes since the FDA cleared the first device in 2008. At that time, it occupied an uncomfortable niche, as the FDA only cleared it for use after failure of one (but not two) antidepressants. Few patients, and even fewer insurers, were willing to take on this costly and time-consuming treatment for such a mild degree of treatment resistance.

Evidence in Depression

Since 2008, TMS has evolved in ways that amplify its therapeutic potency. First came Brainsway's Deep TMS, which stimulates a broader and deeper area of the brain, earning TMS its first approval for true treatment resistance in 2013. Next came intermittent theta-burst stimulation (iTBS) in 2018. These devices deliver the magnetic pulses at a frequency that matches the natural electromagnetic activity in the hippocampus. At first, iTBS was seen as an advance of efficiency, shortening the treatment sessions from 40 minutes to 3 minutes. However, iTBS opened the door to the next advance by allowing clinicians to deliver multiple treatments in the same day, known as accelerated iTBS.

Accelerated iTBS led to faster responses and higher remission rates. Working at Stanford University, Nolan Williams and colleagues brought 90% of patients with treatment-resistant depression to remission by delivering 10 treatments of iTBS a day over an intensive five-day course, spacing the treatments out every hour. By comparison, traditional TMS brings only 20%–35% to remission. When they replicated that success in a randomized controlled trial, the FDA cleared the protocol as Stanford Neuromodulation Therapy (SNT) in 2022 (brand name SAINT).

The trial was remarkable for its high level of remission (79% for SNT vs 13% with sham), and for the fact that participants were not able to tell whether they received the real or sham therapy (Cole EJ et al, *Am J Psychiatry* 2022;179:132–41). Maintaining the integrity of the blind is difficult with powerful, fast acting therapies. The blind failed in trials of psychedelics and ketamine, where most patients accurately guessed their assigned treatment.

SNT owes part of its success to more precise placement of the magnet. TMS affects only a dime-sized area of the brain, about 0.8% of the brain's surface. This localization limits its side effects but can also reduce its efficacy if the magnet misses the target. In depression, that target is the dorsolateral prefrontal cortex (DLPFC), and most devices find it by mapping the thumb. The magnet is moved until the patient's thumb twitches. The DLPFC is approximately 5 centimeters away from the area that controls the thumb. The problem is that these coordinates vary with the shape of the head, so thumb-mapping misses the mark up 30% of the time. To improve on this accuracy, some machines use EEG-guided mapping. SNT takes it a step further by mapping the DLPFC with functional magnetic resonance imaging (Deng ZD et al, *Biomedicines* 2023;11(8):2320).

Only one device is cleared for the SNT protocol, and centers that offer it are listed on the manufacturer's website (magnusmed.com). Other protocols for accelerated iTBS exist that deliver fewer sessions per day or use EEG-guided mapping, but these have not led to remission rates as impressive as those for SNT (eg, 10%–25% vs 60%–90%) (Neuteboom D, *Psychiatry Res* 2023;327:115429).

The next advance on the horizon is magnetic seizure therapy (MST), which repurposes the TMS magnet to induce a therapeutic seizure, much like ECT. MST is not commercially available, but in research settings, it achieved the same efficacy as ECT but with fewer cognitive side effects. It spares cognition by inducing localized seizures that affect only 20% of the brain, while ECT affects 95% (Deng ZD et al, *JAMA Psychiatry* 2024;81(3):240–249).

When to Consider TMS

TMS is FDA cleared in treatment-resistant depression. Its efficacy sits somewhere between pharmacologic augmentation and ECT, and it is better tolerated than both those options. SNT brings the efficacy even higher.

Consider TMS a first-line therapy for difficult-to-treat depressions that do not have psychotic or catatonic features (for those, ECT is more effective).

TMS is approved down to age 15 in depression. Unlike antidepressants, TMS is effective in vascular depression. It poses no known risks during pregnancy and breastfeeding.

TMS is also FDA cleared for obsessive-compulsive disorder (OCD), nicotine cessation, and migraines, although the magnet is aimed at a different part of the brain in those protocols. However, most comorbidities improve when depression is treated with the DLPFC protocol, particularly anxiety disorders (Thompson L, *J Affect Disord* 2020;276:453–460).

The main strikes against TMS are cost and convenience. Although most insurers cover it, patients who have a high deductible will need to pay from $9,000 to $12,000 for a six-week course. Unlike ECT, patients can work while undergoing TMS, but they may need to secure their Family Medical Leave Act (FMLA) rights to leave work each day for the treatment.

Mechanism of Action

If you lifted a twelve-pound dumbbell for 20 minutes, little would change. However, if you did that five days a week for six weeks, you'd see a gradual increase in muscle size and strength. That is a rough analogy for what happens during a course of TMS. Depression disrupts the connections between the parts of the brain that regulate emotions (limbic system) and executive functions (left dorsolateral prefrontal cortex, DLPFC), and TMS strengthens those circuits. TMS also reshapes the part of the brain involved in ruminative thinking, the default mode network (Peng Z et al, *Shanghai Arch Psychiatry* 2018;30(2):84–92).

This remodeling is known as neuroplasticity. Aerobic exercise, psychotherapy, and social interactions also bring about neuroplastic changes in the brain, as does learning to juggle or play piano. In theory, these kinds of neuroplastic habits will enhance the benefits of TMS, and there is some evidence to support that. TMS lights the neuroplastic spark by activating neurons in the DLPFC. The magnetic field repolarizes those neurons, causing them to fire repeatedly during treatment. Along with these neuroplastic changes are increases in brain-derived neurotrophic factor (BDNF), hippocampal neurogenesis, and changes in synaptic transmission. TMS also modulates dopamine, glutamate, and serotonin (specifically, it increases serotonin 5-HT2A receptor binding).

Patients with higher levels of treatment resistance need more magnetic pulses to achieve an antidepressant response, and this finding inspired the repeated sessions in SNT. The frequency of those sessions is also important. In SNT they are spaced 50 minutes apart, as that seems to be the "sweet spot" to produce a cumulative effect on synaptic strengthening.

How to Recommend TMS

Treatment Options and Logistics

TMS is available at neuromodulation centers, academic institutions, and some private psychiatric offices. Outside of SNT, the three TMS protocols are equally effective but differ in duration. Treatment sessions last 3 minutes for iTBS, 20 minutes for Deep TMS, and 20–40 minutes for traditional TMS.

Augmentation Strategies

The way that TMS affects the brain depends on the patient's state of mind when it is delivered. In OCD and post-traumatic stress disorder (PTSD), TMS works better when patients think about their anxiety or obsessions beforehand. With depression, it is the opposite. Directing patients to ruminate on depressing problems diminishes its benefits, while asking them to engage in a more active, problem-solving frame of mind may help (Sack AT et al, *Biol Psychiatry* 2024;95(6):536–544). That isn't easy to achieve during depression, but patients can walk in nature, listen to their favorite music, read an inspiring book, or play a simple game before treatment.

Other ways to augment TMS include:

- Mindfulness meditation
- Psychotherapy
- Morning light therapy
- D-cycloserine (Seromycin)

Treatment with d-cycloserine before the TMS session augments both its clinical and neuroplastic effects (see Chapter 25: D-Cycloserine), but not all treatments are additive. Benzodiazepines and antipsychotics may dampen the response to TMS (Kochanowski B et al, *Harv Rev Psychiatry* 2024;32(3):77–95; Barbini B et al, *Int J Psychiatry Clin Pract* 2021;25(4):375–377).

Maintaining the Response

About half of patients who recover with TMS will relapse within 6–12 months, so some form of prevention is in order. Unfortunately, we don't have as many well-researched options here as we do with ECT. Maintenance TMS lowers the risk of relapse, particularly during the first six months, when the relapse risk is highest (d'Andrea G et al, *J Pers Med* 2023;13(4):697). The treatment is typically lowered gradually over several months, first from five days a week to three days a week, then two days a week, and finally to monthly treatments.

In practice, finances are often a limiting factor, as many insurers do not cover maintenance TMS. Often, a rescue approach is used where TMS is delivered if symptoms arise. Alternatively, patients can repeat a second course of TMS if another episode occurs. More than 80% of patients respond to a second course after a successful first course of TMS, and the second course often requires fewer sessions (Janicak PG et al, *Brain Stimul* 2010;3(4):187–199).

What about those 70%–90% remission rates for SNT? After completing the intensive five-day course, most patients stay well for the first month, but nearly half relapse by month three (Geoly AD et al, *Neuromedic J* 2024;4:2). SNT's manufacturer is developing a protocol to prevent depression by responding to early signs of relapse with as-needed sessions. In a one-year trial, all 21 patients avoided depressive episodes with this approach and required an average of 1 day of SNT therapy per month (Stimpson K et al, *Brain Stimul* 2025;18:208e617).

Side Effects

Tolerability

TMS is better tolerated than most medications, with rates of discontinuation due to adverse events of 5% vs 25% for antidepressants. The most common side effects are headache and scalp discomfort, which affect around 1 in 5 patients. Those side effects are more intense during iTBS, where the magnetic frequency is five times higher.

Risks

TMS is generally safe, with a rare risk of seizures (the risk is 1 in 30,000 sessions; by comparison, the seizure risk with antidepressants is approximately 100 times higher).

Key Takeaways

- TMS is more effective and better tolerated than augmentation, but cost and convenience limit its use.
- Stanford Neuromodulation Therapy (SNT, SAINT) is an intensive, five-day TMS program that brings 70%–90% of patients with highly resistant depression into remission.
- Unlike ECT, TMS preserves or improves cognitive function, but it lacks ECT's specificity in psychotic and catatonic depression.

CHAPTER 43

Electroconvulsive Therapy (ECT)

ECT TREATS DEPRESSION BY INDUCING a seizure, or convulsion, in the brain. Its origins date back to the 1930s, when Hungarian psychiatrist Ladislas Meduna noticed that psychotic symptoms improved after a seizure. Building on this chance observation, Meduna used camphor to chemically induce a seizure in a man with severe catatonia. The patient suddenly began walking, talking, and eating independently for the first time in four years. More success followed, and convulsive therapy spread internationally. In 1938, Ugo Cerletti and Lucio Bini in Rome found a safer method to induce convulsions with electricity, and ECT was born.

Since then, the safety of ECT has continued to evolve. Anesthesia and muscle relaxants arrived in the late 1940s. By limiting the seizure activity to the brain, these advances prevented fractures and brought the mortality rates in line with that of a dental anesthesia. In recent decades, psychiatrists have reduced the cognitive side effects of ECT by applying the current as ultra-brief pulses or limiting it to the right side of the brain (right unilateral ECT). These innovations reduce the efficacy a little but improve the tolerability a lot.

These are welcome advances, but the public perception of ECT remains mired in the past. This is unfortunate, as ECT is the most effective therapy we have for severe depression. ECT lacks the promotional support of the pharmaceutical industry, and it has a powerful adversary in the Church of Scientology (see sidebar).

Evidence in Depression

ECT is the gold standard for psychotic and catatonic depression, with recovery rates around 90% in those cases. ECT is also effective for treatment-resistant depression, schizoaffective disorder, bipolar depression, and delirious mania.

When to Consider ECT

Consider ECT as a first-line treatment for psychotic and catatonic depression. For other depressions, consider ECT when two or more antidepressants have failed and the symptoms are moderate to severe, or when rapid intervention is needed for factors like acute suicidality.

Mechanism of Action

During ECT, electrical stimulation over the cortex produces a generalized, tonic-clonic seizure. Exactly how that seizure treats depression is not fully understood, but probably stems from a combination of changes in brain structure, connectivity, and synaptic signaling. These changes mirror those brought about by antidepressants, but they occur more rapidly and robustly with ECT.

Structurally, ECT increases gray matter volume, particularly in the hippocampus and amygdala, where it is associated with neurogenesis. It reduces neuroinflammation and alters neural networks involved in mood (thalamocortical and corticolimbic circuits). At the neurochemical level, ECT modulates serotonin, dopamine, and norepinephrine, and balances stress hormones like cortisol in the hypothalamic-pituitary-adrenal (HPA) axis (Belge JB et al, *Curr Top Behav Neurosci* 2024;66:279–295).

How to Recommend ECT

Start with a compassionate description of ECT's efficacy. "Your suffering is great. There is a treatment that can work quickly and allow you to live as you did before this depression. That treatment is ECT, and of all the options it has the highest chance of success." Next, give a clear account of the side effects and risks. If the patient is hesitant, advise them to view the visit with the ECT center as a consultation. Even if the patient declines the treatment, the ECT service might have other ideas for their treatment.

ECT is typically delivered three days a week for three to four weeks. Each session lasts at least an hour. The patient is given light anesthesia, rendering them unconscious during the 30–60 second seizure. Patients cannot drive on the day of ECT, and they probably will not be able to work during the entire course because of the cognitive side effects.

Maintaining the Benefits

ECT is powerful and fast, but it does not last. Without something to keep them well, more than 80% of patients relapse within six months of successful ECT. Those rates improve by ending the treatment with a slow taper. After completing the three- to four-week course, the patient receives another four to five sessions tapered over a month (McCall WV, et al, *J Psychiatr Res* 2018;97:65–69). Beyond this taper, there are four other ways to reduce the relapse risk:

- Nortriptyline (Pamelor) + lithium
- Cognitive behavioral therapy (CBT)
- Maintenance ECT

The first strategy comes from a landmark trial where a nortriptyline lithium combination cut the relapse rates in half and worked better than nortriptyline alone (Sackeim HA et al, *JAMA* 2001;285(10):1299–1307). For patients with bipolar depression, lithium can be used as monotherapy, and for patients who cannot tolerate nortriptyline, a serotonin-norepinephrine reuptake inhibitor (SNRI) can be substituted.

CBT's preventative effects are based on a small study where the psychotherapy was combined with medication after ECT. CBT doubled the remission rates compared to maintenance ECT and to medication alone (Brakemeier E, *Biol Psychiatry* 2014;76(3):194–202).

In maintenance ECT, the ECT sessions are continued every two to four weeks for six months and then tapered off, depending on response.

Perhaps the best approach is to combine several of these strategies. For example, nearly all the patients who relapsed on lithium did so in the first month. These relapses might have been avoided if the ECT had been tapered off in the final month while lithium was started. That possibility is untested, but one study found that combining the nortriptyline-lithium strategy with maintenance ECT was twice as effective as either alone (Nordenskjöld A, *J ECT* 2013;29(2):86–92).

Side Effects

Tolerability

Patients may wake up from the treatment with a headache, muscle soreness, and brief confusion. These usually improve with acetaminophen or nonsteroidal anti-inflammatory drugs (NSAIDs). More concerning are the memory problems, which come in two forms.

- Anterograde amnesia, where their ability to store new memories and learn new information is impaired. This problem is temporary. A few weeks after treatment, most patients have better cognition than they did before ECT (Semkovska M and McLoughlin DM, *Biol Psychiatry* 2010; 68(6):568–577). In rare cases, the cognitive side effects last longer, and it is not clear if this is due to residual effects of the depression or to ECT.
- Retrograde amnesia, where they forget recent events that happened before starting ECT. These memories are unlikely to return, but most patients do not regret losing the memory of their depression. Rarely, patients report forgetting events that took place in the distant past, such as a vacation they took or a movie they saw.

Other aspects of memory are spared. Patients do not forget the names of their loved ones, how to drive a car, or how to perform their jobs. ECT does not increase the risk of dementia.

Risks

The main medical risk with ECT is cardiovascular, as the treatment can raise blood pressure and cause arrhythmias. Its adversaries argue that ECT is "deadly," citing data that does not account for the elevated risk of death with severe depression. ECT is likely associated with a very small mortality risk (around 1:50,000), but on balance the treatment lowers the risk of death when compared to patients with severe depression who did not receive ECT (Watts BV et al, *Br J Psychiatry* 2021;219(5):588–593).

Key Takeaways

- ECT is one of the most effective treatment for severe depression, particularly for psychotic and catatonic presentations, with remission rates approaching 90%.
- Modern developments like anesthesia, muscle relaxants, and right unilateral placement have improved ECT's safety and reduced its cognitive side effects.
- ECT causes two memory problems. First is a temporary difficulty with learning new information during the treatment. Second, there is a lasting amnesia for events around the time of ECT.

- Without prevention, 80% of patients relapse after ECT. Strategies to reduce that risk include nortriptyline-lithium combination, tapering ECT off over a few weeks, monthly maintenance ECT, and CBT.

ECT, Brain Damage, and Scientology

Before founding the Church of Scientology, L. Ron Hubbard was a best-selling science fiction writer with a long history of psychiatric treatment. At first, his relationship with psychiatry was positive, inspiring him to volunteer on the wards as a kind of peer support specialist. In the late 1940s, Hubbard and his wife, Sarah, created Dianetics, a system of psychotherapy that mixed psychoanalytic ideas with an electrical device to home in on traumatic memories.

Hubbard hoped to launch a career as a psychotherapist from this work. Though lacking in professional training, his hopes were not without precedent. Melanie Klein developed play therapy and object-relations theory without even a college degree.

Hubbard unleashed Dianetics to the public in a 1950 book that met with swift sales and professional ridicule. As journals from *JAMA* to *Consumer Reports* rejected his claims, the stress started to take a toll. His wife sought the help of a psychiatrist for what she described as increasingly erratic and violent behavior in her husband. The psychiatrist suggested – without meeting Hubbard – that he suffered from schizophrenia. When the informal diagnosis was leaked to the press, the couple divorced and L. Ron Hubbard forever fixed his cannon against psychiatry (note: never diagnose a person without meeting them, and even then, with caution).

Unable to make professional inroads, Hubbard shifted to religion. In 1954, he founded the Church of Scientology, anointing it with a mission to spread Dianetics and eradicate psychiatry. For the later mission, he created the Citizens Commission on Human Rights, and set ECT as their first target. Scientologists recruit patients to sue manufacturers of ECT devices and the psychiatrists who use them. They pressure politicians to outlaw ECT, painting the treatment as a torture device that causes brain damage (Kent SA and Manca TA, *Ment Health Relig Cult* 2014;17(1):1–23).

ECT can cause cognitive problems, but it does not cause brain damage or, to use the medical term, neuronal injury. Brain volumes do not shrink and markers of neuronal injury do not rise after ECT. Instead, we see the opposite. Neuroprotective factors like brain-derived neurotrophic factor (BDNF) rise, and brain volumes increase, a change that is detectable for several months after ECT and is not due to cerebral edema.

Opponents have argued that ECT cooks the brain with electricity, but the temperature change during ECT is within the range of normal physiology (Swartz C et al, *J ECT* 2023;39(3):158–160). ECT does not raise the risk of dementia, and autopsies have revealed no neuronal injury even after long-term ECT (up to 1,250 treatments) (Swartz CM, *J ECT* 2024;40(2):72–77).

Scientology may not inspire public trust, but its attacks on ECT have a way of spreading independent of their attribution. The majority of the public (57%) believe that ECT is harmful. Patient polls, however, tell a different story. Among those who have undergone ECT, 70% found it helpful and say they would do it again (Lauber C et al, *Psychiatry Res* 2005;134(2):205–209; Chakrabarti S et al, *World J Biol Psychiatry* 2010;11(3):525–37).

Other Treatments

CHAPTER 44

Reproductive Hormones

AS PEOPLE AGE, REPRODUCTIVE HORMONES fall: estrogen in women, and testosterone in both genders. These declines can contribute to depression, particularly for women when reproductive hormones are changing rapidly in the transition to menopause.

Mechanism and Background

Reproductive hormones influence neurotransmitter systems involved in mood regulation, including serotonin, dopamine, and norepinephrine. They also affect neurogenesis, inflammation, and the hypothalamic-pituitary-adrenal (HPA) axis—all of which play a causal role in depression.

The relationship between hormones and depression varies by gender and age. In women, abrupt hormonal fluctuations (such as during perimenopause) pose a greater depression risk than consistently low levels (such as in established menopause). In men, the evidence linking testosterone decline to depression is weaker, though a subset of older men with low testosterone may experience mood benefits from replacement.

The decline in hormones can stem from normal aging or from pathological hypogonadism, where a disease state causes the ovaries or testes to stop producing adequate hormone levels. Hypogonadism is classified as primary (originating in the gonads) or secondary (originating in the pituitary or hypothalamus). Several psychiatric conditions can cause or contribute to hypogonadism in men and women:

- Anorexia nervosa
- Use of opioids or anabolic steroids
- Alcohol use disorder
- Hyperprolactinemia, such as from antipsychotics
- Traumatic brain injury

Diagnosing hypogonadism and distinguishing it from normal aging is complex and typically requires specialist evaluation. However, we should recognize when hormonal factors might be contributing to treatment-resistant depression.

Estrogen and Depression in Women

Women are more vulnerable to depression as they transition to menopause. The transition begins with perimenopause, when hormone levels decline and menstrual cycles become less frequent and more irregular. Perimenopause typically begins in the late 40s and lasts four to seven years, although for some women it can start in the 30s. It is followed by menopause, which is defined as a full year without menses. In menopause, levels of estrogen, progesterone, and testosterone are stable but low.

As menopause approaches, women may develop premenstrual mood symptoms even if they didn't have them before. They feel worse one to two weeks before their periods, with depression, irritability, insomnia, and anxiety, as well as physical symptoms like night sweats and hot flashes. The symptoms are more severe when menopause is brought on abruptly through ovariectomy or by taking estrogen-blocking treatments for breast cancer like tamoxifen.

Treatment

Selective serotonin reuptake inhibitors (SSRIs) and serotonin-norepinephrine reuptake inhibitors (SNRIs) are first-line interventions for both the physical and psychological symptoms of perimenopause. Table 44.1 lists antidepressants with the most research, but in practice, it's best to go with one that worked well for the patient before. Other serotonergics like vortioxetine are also effective in this phase, although nonserotonergics like bupropion are probably not.

TABLE 44-1. Treatment of Menopausal Symptoms

SSRIs	For mood and vasomotor symptoms Paroxetine (7.5–15 mg QHS), escitalopram (10–20 mg QHS), citalopram (10–20 mg QHS)
SNRIs	For mood and vasomotor symptoms Desvenlafaxine (50–150 mg QHS), venlafaxine XR (37.5–150 mg QHS)
Gabapentin	For insomnia and vasomotor symptoms (300-2,400 mg at night)

TABLE 44-2. Contraindications to HRT

History of estrogen-sensitive cancer (eg, breast, ovarian, or uterine)
Unexplained vaginal bleeding
History of deep venous thrombosis (DVT), pulmonary embolism, blood clotting disorder, or stroke
Conditions that increase the risk of stroke (uncontrolled hypertension or elevated triglycerides)

Estrogen supplementation can be helpful here and is more beneficial during the perimenopausal period when hormone levels are declining than it is when they are steady and low. In the early phase of menopause, oral contraceptives are often used. As women get closer to menopause, hormone replacement therapies (HRT) are preferred. These have lower hormone levels than oral contraceptives and do not protect against pregnancy. Women who have gone through surgical menopause can also take HRT, as long as they do not have a history of estrogen-sensitive cancers, such as breast, endometrial, or ovarian cancer.

OBGYN physicians often prescribe HRT for premenopausal symptoms like hot flashes, insomnia, and night sweats, but they rarely use it for full depressive episodes. Compared to antidepressants, the evidence that HRT is effective in perimenopausal depression is smaller and the risks are higher, so it is often reserved for women who cannot tolerate or do not respond to antidepressants.

Most of the studies on HRT and mood involved healthy women without clinical depression. There, HRT improves well-being, energy, and cognition, as well as hot flashes, night sweats, weight gain, and vaginal dryness. It reduces the risk of diabetes and osteoporosis but increases the risk of stroke and venous thromboembolism.

The risks of cancer and heart disease are more controversial. The Women's Health Initiative Study found an increased risk of breast cancer and heart disease with HRT. The study was widely publicized when it came out in 2002, but many of its findings were later refuted, with some studies finding that HRT protects the heart (Levy B and Simon JA, *Obstet Gynecol* 2024;144(1):12–23).

When it comes to clinical depression, we only have two randomized controlled trials of HRT in perimenopausal women. Both used HRT as monotherapy. Both were positive, with response and remission rates of

65%–70%, but with only 84 trial participants the uncertainty here is high (Schmidt PJ et al, *Am J Obstet Gynecol* 2000;183(2):414–420; Soares CN et al, *Arch Gen Psychiatry* 2001;58(6):529–534).

These studies used transdermal HRT, which provides steadier serum levels than oral formulations. To protect against endometrial cancer, HRT is given with both estrogen and progestin. Estrogen has mood elevating effects, while progestin can be depressogenic.

HRT is best managed by a specialist, who can better weigh its medical risks and monitor for excessive bleeding, fibroid growth, and blood pressure changes. If you think HRT is indicated, contact the OBGYN or other specialist to explain that other psychiatric treatments have not worked and ask if HRT is a possibility. To minimize the risks, they will often use HRT for a limited time, usually five years, although longer periods are becoming more common as more reassuring data has appeared.

Testosterone and Depression

Testosterone levels decline in both genders, starting around age 30 for men and in the mid-40s for women. While it seems intuitive that this decline is linked to depression, the case is not so clear. Most studies in men find no association, and in women the link is even weaker (Forbes M et al, *J Gerontol A Biol Sci Med Sci* 2025;80(6):glaf019; Hemachandra C et al, *Maturitas* 2023;168:62–70).

The testosterone-depression link is stronger for older men with low-grade depression, the type that was known as dysthymic disorder in DSM-IV. Most cases of dysthymic disorder begin in early childhood, but a subset starts later in life in men, and this subset is linked to low testosterone. Replacement therapy is likely to improve mood in these men, but there are some risks to navigate.

Testosterone is only approved for men whose low levels are due to a medical disorder, such as from chemotherapy, hypothalamic or pituitary disorders, and Klinefelter's disease. However, most prescriptions are written off-label, for age-related deficiencies, and the FDA became concerned when small trials suggested this practice came with health risks, particularly for the heart. In 2015, they ordered testosterone manufacturers to fund the large TRAVERSE trial to test its safety over two years. Depressive symptoms were a secondary outcome in this landmark study, which tested

TABLE 44-3. Estrogen in Psychiatry (OCP and HRT)

OCP vs HRT	Both oral contraceptive pills (OCPs) and hormone replacement therapy (HRT) contain estrogen along with progestin or a synthetic progestin (eg levonorgestrel, desogestrel, drospirenone). HRT contains lower doses that supplement a woman's hormones, while OCP has higher doses that shut down the natural hormones.
HRT for depression	Menopausal depression studies used 17β-estradiol (Estraderm) transdermal 0.05 mg/day with additional medroxyprogesterone acetate 10 mg/day for 1 week of every month
Types of OCP	Monophasic formulations deliver 3 weeks of estrogen/progestin at a constant level followed by 1 week of placebo. Triphasic formulations deliver varying hormones levels throughout the 4-week cycle that more closely mimic the natural cycle. Progestin-only pills are used by women who cannot take estrogen (eg due to cardiovascular or cancer risks or side effects like nausea, breast tenderness, fluid retention, and rash).
OCP most likely to disrupt mood	Progestin formulations, followed by levonorgestrel formulations
OCP least likely to disrupt mood	Desogestrel formulations (eg Yasmin, Yaz)
Risk factors for mood disruption on OCP	Adolescents and women in the postpartum period. Women with a history of perinatal depression or dysmenorrhea.
Estrogen-induced drug interactions	Estrogen-containing OCPs lower lamotrigine by 30%-50%. HRT lowers lamotrigine to a lesser extent. Milder interactions with OCPs include lower levels of benzodiazepines metabolized through glucuronidation (lorazepam, oxazepam, temazepam), higher levels of other benzodiazepines (alprazolam, chlordiazepoxide, diazepam, flurazepam, triazolam), and higher levels of some tricyclic antidepressants.
Drugs that lower OCP and HRT levels	Carbamazepine (very potent), oxcarbazepine, topiramate, the modafinils, and St John's Wort (through CYP-3A4 inhibition). Transdermal and implant formulations partially bypass this hepatic interaction but are still affected by it. If prescribing these meds, recommend barrier methods and/or higher estrogen doses.

TABLE 44-4. Risks of Testosterone Supplementation

Risks of Testosterone Supplementation
Atrial fibrillation
Acute kidney injury
Thrombosis (eg, pulmonary embolism)
Benign prostatic hyperplasia (BPH)
Low sperm count
Worsening of sleep apnea
Hyperlipidemia
Elevated liver function tests
Acne
Aggression

transdermal testosterone in 5,204 men age 45 to 80 years (Bhasin S et al, *J Clin Endocrinol Metab* 2024;109(7):1814–1826).

The study confirmed what we knew from earlier trials, that testosterone improved mood and sexual functioning in older men whose levels were low. In terms of safety, the results were reassuring in some respects but not others. Testosterone replacement did not raise the risk of heart attacks, stroke, or prostate cancer. On the other hand, the study identified new risks of atrial fibrillation and acute kidney injury and confirmed the known risk of pulmonary embolism. However, these risks were low, so the overall effect of the trial was to ease the path for testosterone supplementation, at least in those who met the study's inclusion criteria.

Those criteria are worth knowing as they clarify when testosterone therapy is appropriate. The men had to have two fasting low testosterone levels (< 300 ng/dL) and at least one symptom of hypogonadism:

- Low or depressed mood
- Fatigue or low energy
- Low sex drive
- Decreased spontaneous erections
- Loss of axillary or pubic hair or reduced need to shave
- Hot flashes

While testosterone may improve mild depression in older men with low levels, it has failed in other populations, including men with low levels and full depression and in women with depression. Supplementation can cause aggression, as well as medical risks that are listed in Table 44.4, but it also has medical benefits. For men with low testosterone, supplementation can improve libido, erectile function, anemia, bone mass, and glycemic control.

Oral Contraceptives, Teens, and Depression

Reproductive hormones can have opposite effects in older and younger women. HRT improves mood when estrogen declines during perimenopause, but those hormones can worsen mood when used as contraception in younger women with normal estrogen levels. The risk is small, but we don't know which younger women are more at risk.

The oral contraceptives that are least likely to disrupt mood are those that contain drospirenone and ethinyl estradiol (eg, Yaz and Yasmin) (Kelly S et al, *Clin Drug Investig* 2010;30(5):325-336). Intrauterine devices are also safer, as they contain lower hormone levels. Riskier versions include the triphasic preparations, where the hormone levels fluctuate more, and progestin and levonorgestrel formulations.

Another way to reduce this risk is to space out the periods by skipping the placebo week that allows menstrual flow. This strategy relieves premenstrual symptoms, both physical and mental, by triggering menstruation every six months instead of every month. It is best done under the guidance of the clinician prescribing the oral contraceptive.

Key Takeaways

- In women, hormone replacement therapy (HRT) may improve depressive symptoms in the perimenopausal period.
- Oral contraceptive pills can worsen mood in younger women. This risk is lower with formulations that contain drospirenone and ethinyl estradiol (eg, Yaz and Yasmin).
- Some older men develop a low-grade depression as their testosterone declines, and testosterone may improve mood if their level falls below 300 ng/dL.
- HRT and testosterone carry medical risks and require monitoring that is best handled by a specialist.

Deprescribing

CHAPTER 45

Overview of Deprescribing

WHEN PATIENTS WITH DIFFICULT-TO-TREAT DEPRESSION arrive at your door, they're often lugging a pharmacologic suitcase packed with medications that bear little resemblance to the evidence-based strategies in this book. The typical regimen—multiple antidepressants, an antipsychotic with a stimulant, a benzodiazepine, and maybe an anticonvulsant—reads like a history of psychiatric fads rather than a coherent treatment plan. Meanwhile, the therapies with the strongest evidence in treatment resistance—lithium, pramipexole, thyroid, monoamine oxidase inhibitors (MAOIs), electroconvulsive therapy (ECT), and transcranial magnetic stimulation (TMS)—are notably absent.

Deprescribing isn't just about clearing the plate. It's about recognizing that each additional medication carries diminishing returns and mounting risks. Polypharmacy causes drug interactions, side effect burden, and medication errors. It also complicates the diagnostic picture—is that insomnia a symptom of depression, or a side effect of the stimulant? Is the anxiety part of the illness, or is it due to akathisia on an antipsychotic?

Before suggesting any changes to a complex regimen, we need to address three key questions about each medication:

Response

"How did you respond after starting it? Did it work all the way, part of the way, not at all, or unsure?" Look for times off the medication, whether through missed doses or intentional discontinuation. Listen for specifics rather than generalizations like "it helps." Which symptoms improved, and when did the patient notice a change?

Psychological attachment

Patients develop relationships with their medications that transcend pharmacology. They may be psychologically attached even when they don't believe a medication helped, as in "I'm sure I'd be worse off it." They may view your suggestions to taper off as giving up on them or dismissing their suffering.

Scientific evidence

Is there evidence that the medication works in depression? There is thin or contradictory evidence that many commonly prescribed medications are effective in treatment-resistant depression. Some worked in small trials but failed when tested rigorously. Others showed benefits only in specific subgroups of patients.

After gathering that information, make a collaborative decision with the patient about whether to discontinue the medication. In the next chapters I'll sketch out tapering schedules, but tapering is a rough science. Building a strong therapeutic alliance does more to ease withdrawal than finding the precise dosing increment.

Timing Matters

Deprescribing requires strategic timing. It's more successful when the patient's life and health are relatively stable. Attempting to taper medications during a period of significant stress, physical illness, or major life transition is rarely successful.

If a taper doesn't work but there's still good reason to come off, I often try again after one to two years. About half the time, the second attempt succeeds. What changes? Sometimes the patient's circumstances have improved, sometimes their illness has receded, and sometimes they're simply more prepared psychologically.

Building Resilience Before Tapering

Equally important is the patient's own resilience. Tapering is a good time to build in antidepressant habits—mindfulness, exercise, the Mediterranean diet, regular sleep and wake times. These nonpharmacologic interventions not only support mood directly but also buffer against withdrawal symptoms.

I often tell patients, "We're not just taking something away; we're making room for something better." That might mean an evidence-based medication strategy, or it might mean the mental clarity to engage more fully in psychotherapy.

In the pages that follow, I'll walk through medications to consider deprescribing when they show up in complex regimens for treatment-resistant depression. For each, I'll review the evidence, outline tapering strategies, and help you identify the patients who might benefit from continued treatment despite the thin evidence base.

Clinical Vignette

Brittany arrived with a list of eight psychiatric medications, including three antidepressants, two benzodiazepines, a stimulant, an antipsychotic, and an anticonvulsant. When we reviewed each one, it became clear that she couldn't identify benefits from most of them. "I'm not sure which ones are working," she admitted, "but I'm afraid to stop any of them." We started with the medication that seemed to cause the most side effects—the antipsychotic, which gave her akathisia. After successfully tapering that, Brittany gained confidence. Over two years, we gradually reduced her regimen to two evidence-based medications (an MAOI and lithium), which provided better symptom control than her previous eight-drug cocktail.

Withdrawal Symptoms to Monitor

Table 45.1 outlines common withdrawal symptoms by medication class. Before tapering, review these with patients so they can distinguish temporary withdrawal effects from from true depressive relapse.

TABLE 45-1. Withdrawal Symptoms by Medication Class

Medication	Withdrawal Symptoms
Antipsychotics	**Psychiatric:** Insomnia, anxiety, psychosis **Physical:** Tremor, headache, dizziness, nausea, tachycardia, akathisia, dyskinesias
Benzodiazepines	**Psychiatric:** Anxiety, insomnia, nightmares, irritability, sensory changes, dissociation, paranoia **Physical:** Tremor, tinnitus, headache, dizziness, muscle spasms, GI distress, palpitations, seizures
Bupropion	**Psychiatric:** Anxiety, insomnia, irritability **Physical:** Fatigue, headache, myalgias, restlessness

TABLE 45-1. Withdrawal Symptoms by Medication Class

Medication	Withdrawal Symptoms
Buspirone and gepirone	None known
Mirtazapine	**Psychiatric:** Depression, anxiety, insomnia **Physical:** Itch, appetite loss
Modafinils	**Psychiatric:** Depression, anxiety, hypersomnia, brain fog **Physical:** fatigue, headaches, sweats, chills
SSRIs and SNRIs	**Psychiatric:** Dysphoria, anxiety, mania, insomnia, irritability, brain zaps, sensory changes, dissociation **Physical:** Tremor, sweats, myalgias, restlessness, headache, chills, GI distress, tinnitus
Stimulants	**Psychiatric:** Depression, irritability, anxiety, insomnia, nightmares, hypersomnia, brain fog, agitation **Physical:** Crashing fatigue, dystonic movements, increased appetite

CHAPTER 46

Deprescribing Extra Antidepressants

ANTIDEPRESSANT POLYPHARMACY IS COMMON in treatment-resistant depression, but the evidence rarely supports these combinations. This chapter walks through the most common redundant antidepressants, examining their evidence base and providing practical guidance for tapering when appropriate. The goal isn't to eliminate medications reflexively, but to simplify regimens in ways that improve both efficacy and tolerability.

SSRI and SNRI antidepressants

There is no theoretical or clinical rationale to combine multiple serotonergic medications, but it happens. This one is low-hanging fruit for deprescribing.

Start by educating the patient about the risks with this class, which are probably greater in combination. Serotonergic antidepressants can cause restless sleep, weight gain, gastric distress, apathy, and sexual dysfunction. Medical risks include falls, gastric bleeding, hyponatremia, osteopenia, and serotonin syndrome.

When to Taper Off SSRIs and SNRIs

There is never an indication to take two of these together, so the only dilemma is choosing which one to taper. If the patient is unsure, choose one that is more likely to cause side effects, drug interactions, or medical risks. Among the serotonin-norepinephrine reuptake inhibitors (SNRIs), desvenlafaxine is the safest, as venlafaxine can raise blood pressure and duloxetine can injure the liver. Among the selective serotonin reuptake inhibitors (SSRIs), escitalopram, citalopram, and low-dose sertraline (< 150 mg) do not cause drug interactions; fluoxetine has the lowest risk of weight gain; sertraline is the safest in cardiovascular disease; and paroxetine has the most side effects.

As long as the patient remains on another SSRI or SNRI, withdrawal problems are rare, but it is still best to taper slowly over four to eight weeks. If the patient is tapering off this class completely, a hyperbolic taper minimizes the risk of withdrawal problems. The goal here is to taper down by larger amounts until the minimally effective dose is reached. After that, the dose is reduced in smaller increments, often requiring a compounding pharmacy or a liquid formulation.

Anxiety, insomnia, and depression can be signs of withdrawal, but how do you tell if they represent a new depressive episode? The difference is in the timing. Withdrawal symptoms start within days of a missed dose, peak within a week, and improve quickly when the antidepressant is restarted. A new depressive episode returns more gradually, usually weeks after a taper, and gets better slowly when treatment is started.

A Hyperbolic Taper for SSRIs

1. Lower to the Minimum Effective Dose

Reduce the dose to the minimum suggested in Table 46.1 if not already there (eg, citalopram 20 mg). At this stage, the dose can be reduced linearly (eg, by 5–10 mg) because the main risk is depressive relapse, not serotonin withdrawal. Lowering every two to four weeks is reasonable for most patients, but longer intervals may be needed for those with a history of withdrawal problems or a long duration of treatment.

2. Assess Baseline Symptoms

Check if the patient is having any symptoms that correspond to SSRI withdrawal symptoms at baseline.

3. Lower For One Month and Reassess

Now move to the first tapering dose in the table (eg, citalopram 10 mg). Monitor for withdrawal symptoms throughout the taper and adjust the rate of taper based on withdrawal symptoms.

4. Start the Long-Tail Taper

The doses for each step of the final taper are listed in Table 46.1 (eg, citalopram 5 mg, then 3.4 mg). How quickly you go down at each step depends on how sensitive the patient is to withdrawal. Your assessment of their symptoms at baseline and one month later will give you a sense of that. At a minimum, allow two weeks between each step; four weeks is a rough average. Patients who experience withdrawal effects benefit from making smaller reductions at each step rather than spacing out the time in between dose reductions.

TABLE 46-1. Tapering Doses and Liquid Conversions for Common SSRIs

Medication	Minimum Daily Dose*	Tapering Doses (mg/day)	Liquid Conversion (mL/day)
Citalopram	20 mg	10 mg→ 5→ 3.4→ 2.3→ 1.5→ 0.8→ 0.4→ stop	2 mg/mL: 5 mL→ 2.5→ 1.7→ 1.2→ 0.8→ 0.4→ 0.2→ stop
Escitalopram	10 mg	5 mg→ 2.7→ 1.7→ 1.2→ 0.7→ 0.4→ 0.2→ stop	1 mg/mL: 5 mL→ 2.7→ 1.7→ 1.2→ 0.7→ 0.4→ 0.2→ stop
Fluoxetine	20 mg	8.5 mg→ 4.5→ 2.7→ 1.7→ 1.0→ 0.6→ 0.3→ stop	4 mg/mL: 2.1 mL → 1.1→ 0.7→ 0.4→ 0.3→ 0.2→ 0.1→ stop
Fluvoxamine	50 mg	25 mg→ 15→ 10→ 8→ 5→ 2→ 1→ stop	No liquid (use 25 mg tabs or compounding pharmacy)
Paroxetine	20 mg	11.4 mg→ 7.4→ 5.0→ 3.4→ 2.2→ 1.3→ 0.6→ stop	2 mg/mL: 5.7 mL → 3.7→ 2.5→ 1.7→ 1.1→ 0.7→ 0.3→ stop
Sertraline	50 mg	25 mg→ 14→ 9.1→ 5.9→ 3.8→ 2.2→ 0.9→ stop	20 mg/mL: 1.3 mL → 0.7→ 0.5→ 0.3→ 0.2→ 0.1→ 0.05→ stop

*The minimum daily represents the dose that typically achieves 80% occupancy at the serotonin receptor. It roughly corresponds with the minimum effective dose for depression.

Bupropion

Bupropion inhibits the reuptake of norepinephrine and dopamine, with stronger affinity for the norepinephrine side. That profile has little overlap with that of SSRIs, which has made the combination a popular pairing. Successful case reports and uncontrolled trials led to its inclusion in the STAR*D trial, which brought it more respectability. All that was missing was proof that it worked.

The proof, in the form of randomized trials, has been mixed. In short, bupropion augmentation failed in placebo-controlled trials but appeared to work in trials that compared it to aripiprazole. Aripiprazole is one of the best-studied augmentation agents, so equaling this antipsychotic in two out of three large trials is no small feat. On the other hand, the placebo effect is often just as powerful, so without that comparison it is hard to say what worked in these trials.

The most definitive failure was buried in an unpublished trial from 2009. In this large, industry-sponsored study, bupropion augmentation worked no better than placebo in patients whose depression did not respond to an SSRI (Janik A et al, *JAMA Psychiatry* 2025;82(9):877–887). It was followed by two smaller placebo-controlled trials that delivered equivocal results.

The best we can say about bupropion augmentation is that it might work. In the randomized trials that lacked a placebo, bupropion augmentation did work better than a switch to bupropion monotherapy. In the large CO-MED trial, bupropion augmentation worked better in patients with obesity, inflammation, or mixed features.

One problem with bupropion augmentation is that most SSRIs block the formation of its active metabolite, hydroxybupropion, potentially hindering its antidepressant effects. No studies have tested this idea, but if true it means that bupropion augmentation will be more successful with escitalopram and citalopram, which don't affect the CYP2B6 enzyme that creates this metabolite.

When to Taper Off Bupropion

Bupropion is a good candidate to taper off when the patient is taking multiple antidepressants and its benefits are unclear. Before tapering, conduct a careful review of the pros and cons of each antidepressant the patient is taking. Sometimes bupropion was added to an SSRI that wasn't working. If the SSRI is causing weight gain, apathy, fatigue, or sexual side effects, and bupropion is well tolerated, it may be better to taper the SSRI.

On the other hand, tapering an SSRI or SNRI involves more withdrawal risks than tapering bupropion. Although withdrawal problems are usually mild with bupropion, it is still best to taper it slowly, over two to eight weeks.

Buspirone and Gepirone

Buspirone and gepirone share a similar mechanism but different approvals. Both are agonists at the serotonin 5-HT1A receptor, but buspirone is FDA-approved as Buspar in generalized anxiety disorder (GAD) and gepirone as Exxua in major depressive disorder (MDD).

These medications may work in anxious depression, but beyond that their benefits in depression are uncertain. Gepirone's approval was controversial, with clear benefits in only two of the 15 trials reviewed by the FDA. Gepirone is not known to augment antidepressants, and buspirone failed there, although buspirone did work as monotherapy in anxious depression.

When to Taper Off Buspirone and Gepirone

Consider deprescribing if these medications were started to augment an antidepressant and the patient doesn't have significant anxiety. These medications have no known withdrawal effects, so can be tapered off quickly over a few weeks.

Mirtazapine

In theory, mirtazapine's profile compliments that of serotonergic antidepressants. By blocking 5-HT2 and 5-HT3 receptors, it can reduce serotonergic side effects like nausea, insomnia, akathisia, and sexual dysfunction. These complimentary effects were once thought to enhance antidepressant efficacy. That promise was supported by small early trials and hyped by the nickname "California Rocket Fuel" (coined for the venlafaxine-mirtazapine combo).

That rocket started to falter in 2011 when venlafaxine-mirtazapine proved no better than escitalopram monotherapy in the large CO-MED trial. Seven years later, it crashed, as mirtazapine failed to augment various antidepressants in three large trials involving 796 patients. In one of those studies, switching to a tricyclic (imipramine) was twice as effective as augmenting venlafaxine with mirtazapine.*

Mirtazapine augmentation may still have a role for specific symptoms like insomnia and anxiety. In addition to having sedative qualities, mirtazapine deepens sleep. In one of the failed trials, mirtazapine augmentation did work for a subset of patients with high baseline anxiety.

---

*Careful readers may recall a recent analysis that favored mirtazapine augmentation in depression. This result hinged on the inclusion of less reliable, small trials (all six of the large mirtazapine trials in the analysis were negative) (Henssler J et al, *JAMA Psychiatry* 2022;79(4):300–312).

When to Taper Off Mirtazapine

If a patient is taking mirtazapine along with another antidepressant, first try to figure out which of the two medications might be helping and which is causing side effects. If mirtazapine does not seem to be helping, taper off over two to four weeks. Withdrawal problems are rare.

Key Takeaways

- Multiple antidepressants are rarely more effective than one.
- Tapering off an SSRI or SNRI can create a withdrawal syndrome that resembles an anxious or mixed depression, and a slow hyperbolic taper helps avoid that.
- Three common augmentation strategies—bupropion, mirtazapine, and buspirone—failed in large trials but may have benefits in specific subtypes.

CHAPTER 47

Deprescribing Benzodiazepines, Stimulants, and Anticonvulsants

WHEN DEPRESSION DOES NOT RESOLVE, clinicians often turn to symptomatic treatments like stimulants for fatigue, benzodiazepines for anxiety, or anticonvulsants for irritability. Over time, these complicate the clinical picture and rarely address the core pathology. This chapter examines the evidence behind these common add-ons and provides strategies for tapering when appropriate.

Benzodiazepines

In depression, benzodiazepines have short-term benefits but long-term harms. As we saw in Chapter 28: Alprazolam and the Benzodiazepines, they gradually wear down motivation, cognition, and problem-solving abilities. Still, around 50% of patients with treatment-resistant depression take a benzodiazepine. Discontinuation is promising but risky. Some studies suggest higher mortality rates after coming off a benzodiazepine, while others suggest better mood and cognition off them (Maust DT et al, *JAMA Netw Open* 2023, 6(12)).

To minimize withdrawal problems, taper gradually, reducing by 5%–25% every 2–4 weeks. Reduce by larger dose increments in the beginning and smaller increments toward the tail. The longer the patient has been on the benzodiazepine, the longer the taper. Here is a rough guide to tapering based on the duration of treatment:

- Less than 2 months: Taper over 1 week
- 2–6 months: 3–6 weeks
- 6–12 months: 2–3 months
- 1–3 years: 4–6 months
- Over 3 years: 6 months or longer

Some guidelines recommend switching to a long-acting agent like diazepam or chlordiazepoxide for the taper, but patients may have difficulty with the switch, particularly if changing from alprazolam.

There are no proven antidotes for benzodiazepine withdrawal. Small, controlled trials support carbamazepine (200–800 mg/day), pregabalin (200–400 mg/day), and propranolol (60–120 mg/day). Dividing these doses throughout the day can provide psychological relief, enhancing the placebo effect.

Anecdotally, some clinicians have had success with Silexan, a proprietary extract of lavender approved for generalized anxiety disorder (GAD) outside of the US. Silexan has a large effect size in anxiety and equaled low-dose lorazepam and surpassed paroxetine in randomized trials. Its pharmacology is complex, including effects on glutamate, serotonin, and calcium signaling. Like benzodiazepines, it enhances GABAergic transmission but had no rewarding qualities when tested in recreational drug users.

Silexan has dose-dependent effects on anxiety between 80–160 mg/day, divided BID or given all at night. It is available by prescription in Europe, and the manufacturer has licensed an over-the-counter version in the US (CalmAid by Nature's Way).

Cognitive behavioral therapy (CBT) has the best evidence in benzodiazepine withdrawal (Takeshima M et al, *Psychiatry Clin Neurosci* 2021; 75(4):119–127). This version of CBT teaches skills to manage physical symptoms of anxiety like deep breathing and muscle relaxation, and is available as a self-guided book, the *Stopping Anxiety Medication Workbook* by Michael Otto and Mark Pollack (Oxford University Press, 2009).

When to Taper Off Benzodiazepines

If the patient has an anxiety disorder for which benzodiazepines are effective and the medication is not causing problems, consider staying on it. Among the anxiety disorders, benzodiazepines have the most robust evidence in panic and GAD, for which some of them are FDA-approved, followed next by social anxiety disorder. Intermittent use for simple phobias is also appropriate, but they are not effective in post-traumatic stress disorder (PTSD) or obsessive-compulsive disorder (OCD).

Scenarios that might prompt discontinuation include traffic accidents, falls, overdose attempts, cognitive problems, respiratory illness, older age, and opioid use (particularly in high doses; ie, morphine milligram equivalents over 50; check at www.oregonpainguidance.org/opioidmedcalculator).

If there is no urgent need to discontinue the benzodiazepine, and no clear rationale for continuing it, discontinuation is in order when the time is right. This is a difficult procedure. It goes more smoothly when a secure therapeutic relationship is in place.

Stimulants: Amphetamines and Methylphenidate

Long before they were used in ADHD, stimulants were marketed as antidepressants. This dates back to the 1940s, when the American Medical Association (AMA) oversaw approval of medications. The AMA's criteria was loose, however, and prone to bias. Only approved agents could advertise in AMA journals, which means that more approval meant more revenue for the organization. In the 1960s, the Food and Drug Administration (FDA) took over this task, requiring proof of efficacy and safety in placebo-controlled trials. Stimulants did not have that proof, but the FDA grandfathered them in as antidepressants. When no proof emerged, the FDA ordered the manufacturers to stop marketing amphetamines and methylphenidate as antidepressants in the 1970s.

Thirty years later, stimulants reemerged as a treatment for adult ADHD, and the manufacturers of an amphetamine (Vyvanse) and methylphenidate (Concerta) attempted to clear their path to approval in depression. They were tested over 4–8 weeks as augmentation after an antidepressant failure in four placebo-controlled trials involving a total of 2,572 patients. In all four trials, the stimulants failed to make a difference, not after one week, and not at the end.

These studies are at odds with clinical lore, as many patients report they feel and function better on the drugs. Their pharmacologic properties may explain the variance. At first, stimulants increase synaptic dopamine, raising alertness and confidence, and causing a general state of euphoria. When night falls, they worsen sleep, so that energy and concentration are worse the following day. Those symptoms are relieved by a second dose, but over time the excess dopamine causes problems. It is neurotoxic, causing oxidative stress, inflammation, and damaging dopaminergic nerve terminals in the hippocampus and cerebral cortex.

In practice, this means the initial benefits are likely to wear off and the patient will ask for higher doses. These neurotoxic effects are generally seen in high doses of stimulants, but two animal studies have documented the problem in therapeutic doses of amphetamines (equivalent to 40–60 mg/day of Adderall) (Ricaurte GA et al, *J Pharmacol Exp Ther* 2005;315(1):91–98; Melega WP et al, *Brain Res* 1997;766(1–2):113–120). Clinical problems are also more pronounced on amphetamines, which can worsen the course of comorbid disorders like psychosis, mania, borderline personality disorder, anxiety, and OCD.

Methylphenidate is comparatively safer. This stimulant has a mix of neuroprotective and neurotoxic effects, perhaps because it does not flood the nerve terminal with dopamine as aggressively as amphetamine does. Like amphetamines, methylphenidate can cause dysphoria, paranoia, aggression, anxiety, and compulsivity, but the risk is lower, and methylphenidate may improve some comorbidities. For example, it improved OCD as an augmentation to SSRIs, and—in borderline personality disorder with comorbid ADHD—methylphenidate improved executive functioning.

Getting back to those large, negative trials in depression: Even though both stimulants failed to augment antidepressants, methylphenidate at least eked out a few positive but statistically questionable signals. Methylphenidate may also help fatigue and apathy in geriatric patients and those with terminal illnesses like cancer.

When to Taper Off Stimulants

If the patient is taking the stimulant for a legitimate reason like ADHD, it's best to continue. Otherwise, start by bringing the dose into the safer range if not already there. That means a daily dose at or below:

- Methylphenidate 60 mg
- Dexmethylphenidate 30 mg
- Dextroamphetamine 30 mg
- Lisdexamfetamine (Vyvanse) 70 mg
- Mixed amphetamine salts (Adderall) 35 mg

Stimulants tend to cause more problems when they cross those thresholds, and the thresholds are likely to be lower (by 50%–70%) for older adults (Moran LV et al, *Am J Psychiatry* 2024;181(10):901–909; Farhat LC et al, *JAMA Psychiatry* 2024;81(2):157–166).

Since these are controlled substances, you cannot depend entirely on the patient's self-report to gauge their benefits. Look for objective changes in functioning and use a rating scale for depression and ADHD while lowering the dose (eg, the PHQ–9 and ASRS [Adult ADHD Self-Report Scale] are available online). Often the patient will complain that they feel worse as the dose is lowered, while the rating scale is unchanged or even improved. If they are unable to taper off entirely, consider switching to modafinil or armodafinil, which have a more favorable profile in depression.

One of my patients, a 29-year-old graduate student, was taking Adderall 40 mg daily for "depression with low energy." After initiating a slow taper, he complained that he couldn't concentrate on his studies without it. We added a formal ADHD assessment, which was negative. As his dose was reduced by 5 mg monthly, his sleep gradually improved, and his complaints about concentration problems diminished. By the time he reached 15 mg daily, he noted that his mood was more stable and his anxiety had decreased substantially. Eventually, he was able to discontinue completely, using caffeine strategically for occasional energy boosts.

Modafinils

Modafinil, and its isomer armodafinil, are novel stimulants that are FDA-approved to treat excessive sleepiness in three conditions: narcolepsy, obstructive sleep apnea, and shift work disorder. Modafinil also treats childhood ADHD but failed to gain approval there when a child developed Stevens-Johnson Syndrome in the registration trials, a rare but real risk with these drugs (Wang SM et al, *J Psychiatr Res* 2017;84:292–300).

Thus, modafinil's clinical profile is similar to that of traditional stimulants, bringing rapid improvement in alertness and energy. Like traditional stimulants, modafinil can cause anxiety, insomnia, headache, tremor, hypertension, and rare cardiac arrhythmias. However, it has a lower misuse liability and a lower risk of inducing mania and psychosis. Like methylphenidate and amphetamine, it inhibits dopamine reuptake, but unlike the traditional stimulants, modafinil is neuroprotective.

However, modafinil is not a universal energy pill. It often fails in trials of fatigue related to medical illnesses or medications, including clozapine. In the early 2000s, Shire Pharmaceuticals tested modafinil in patients who had continuous fatigue despite partially responding to an SSRI. It did not work, not for fatigue and not for depression, in this well-designed trial.

Modafinil and armodafinil have since fizzled out in trials of bipolar and unipolar depression, with a pattern shared by other medications in this chapter. The better the trial design, the more negative the results. It even failed to lift mood in a large trial of depression with comorbid sleep apnea, although it did live up to its FDA approval there by improving wakefulness.

As with the traditional stimulants, these results challenge clinical experience. Many patients appreciate the modafinils. If they have side effects, they try to stick with it by lowering the dose. If their insurance denies it, they find a pharmacy with a low out-of-pocket cost. While some of this may be due to its mild rewarding qualities, a signal in the trials suggests another explanation.

Although modafinil did not lift fatigue in the large Shire-sponsored trial, it did bring significant improvements in the overall clinical impression or CGI. The reason may have to do with its cognitive benefits. Modafinil improved working memory and executive functioning in open-label and controlled trials of depression, including in patients who suffered from cognitive problems after recovering from depression (Vaccarino SR et al, *J Clin Psychiatry* 2019;80(6):19r12767). These findings are in line with trials in ADHD and healthy subjects, where modafinil improved various cognitive measures including memory, attention, reaction time, reasoning, planning, and decision-making.

When to Taper Off the Modafinils

Consider tapering off if the patient does not have a comorbid disorder for which it is effective (eg, sleep apnea, narcolepsy, shift work disorder, or ADHD).

The modafinils are not as risky as traditional stimulants, so the need to taper them off is less pressing. They are often used for residual symptoms of depression, but those symptoms often improve with time, allowing them to be deprescribed. As patients get more physically active in the day and sleep better at night, they are less likely to need the modafinils. Compared to traditional stimulants, their withdrawal symptoms are similar but less severe. They can usually be tapered off gradually over two to four weeks.

Anticonvulsants

Anticonvulsants do not have enough evidence to recommend them in unipolar depression, but they do have a small potential there that suggests caution before tapering them off. Among them, lamotrigine comes closest to a valid therapeutic effect. In bipolar disorder, lamotrigine has stronger preventative benefits against the depressive than the manic pole, and in epilepsy, it improves mood independent of its effects on seizures.

Eight randomized controlled trials have tested lamotrigine augmentation in 677 patients with treatment-resistant unipolar depression. Although many of the trials were negative, they turned up a positive result when meta-analyzed together (target dose 50–200 mg/day). Lamotrigine worked best in patients with a longer duration of illness (eight or more years) and more severe symptoms (Goh KK et al, *J Psychopharmacol* 2019;33(6): 700–713).

For other anticonvulsants, the benefits in depression are less clear, and what follows comes from small, controlled trials. Carbamazepine may prevent episodes in recurrent depression. Topiramate may improve depression, but it can also worsen mood. Valproate, gabapentin, and pregabalin don't address depression directly but do have anxiolytic effects.

When to Taper Off Anticonvulsants

When anticonvulsants appear on a patient's regimen, it is a sign that one of their past providers suspected bipolar disorder, so look carefully for this diagnosis before deprescribing. For patients without signs of bipolar disorder, consider continuing the anticonvulsant if it improved depression, reduced the frequency of episodes, or improved anxiety or insomnia. Otherwise, the anticonvulsant can usually be tapered off over two to four weeks, often with improvements in cognition and energy.

Last year I saw a 42-year-old woman with recurrent depression who had been taking lamotrigine 200 mg for five years. Her history revealed four major depressive episodes, each lasting 6–9 months, with complete recovery between episodes. She had no signs of bipolarity. However, since starting lamotrigine, she had experienced only mild, brief depressive symptoms that resolved quickly. This pattern suggested lamotrigine was reducing the cycles of depression, so we elected to continue it despite the lack of a bipolar diagnosis.

Key Takeaways

- Around 50% of patients with difficult-to-treat depression take a benzodiazepine, and their mood and cognition may improve off it.
- Despite their popularity, psychostimulants failed in large, well-designed trials, and are best reserved for comorbid ADHD.
- Anticonvulsants are generally not helpful in unipolar depression, but taper with caution as there is a hint of evidence that they may improve mood (lamotrigine and carbamazepine) or anxiety.

Conclusion

CHAPTER 48

Getting Well and Staying Well

WE'VE JUST REVIEWED A DIZZYING LIST of options for difficult-to-treat depression, but do any of them bring full and lasting recovery? Some work acutely, while others are slow to act but bring more lasting benefits. The best strategies thoughtfully combine the two.

We saw that when electroconvulsive therapy (ECT) is followed by an antidepressant-lithium combination, and when Stanford Neuromodulation Therapy (SNT or SAINT TMS) is followed with as-needed TMS treatments. With ketamine, the answer is less clear, but patients might achieve lasting gains by following it with psychotherapy.

Those three interventional therapies are difficult to access, but we can still apply the strategy with everyday treatment. We need to educate patients that treatments that get them well are not necessarily going to keep them well. Those that are stronger on the preventative side include psychotherapy and lithium, and treatments that address lifestyle factors or nutritional deficiencies: diet, exercise, behavioral activation, light therapy, vitamins, omega-3, and probiotics. Preventative treatments have a slow build. They are also slow to wear off, which creates a problem for adherence.

I often see patients who become depressed a few months after stopping psychotherapy, exercise, or lithium. When I point out the connection, they are unmoved, "Yes, but that was three months ago that I stopped exercising. It doesn't explain why I'm depressed now." It is not easy to stick with treatments that have a delayed reward. Patients with depression are quick to give up on tedious tasks, and this trait remains even after recovery (Valton V et al, *bioRxiv* 2025;2024.06.17.599286).

Hippocrates captured the uncertainty of this work in his first aphorism on the art of medicine. "Life is short, the art long, opportunity fleeting, experiment treacherous, judgment difficult." He followed that pessimistic note with solid guidance that still rings true. "The physician must be ready, not only to do his duty himself, but also to secure the cooperation of the patient, of the attendants and of externals" (ie, family).

That is the heart of this work. It is a difficult task, whether convincing a patient to undergo an accelerated course of TMS, to stay active in the absence of rewards, or to stay alive when hope has temporarily vanished.

Appendix

Treatment Menus

TO ENHANCE COLLABORATION, present your patient with a menu of reasonable options. Keep it brief, 3–4 treatment steps that are likely to work. Aim for variety. Instead of presenting three antipsychotics, choose one treatment that is tolerable, one that is very effective, and one from the natural or lifestyle side. In the lists below, I've divided the treatments in this book into approximate categories to get you started.

Favorable Tolerability

- Celecoxib
- Eszopiclone
- Light Therapy
- L-methylfolate
- Omega-3 fatty acids
- Pramipexole
- Probiotics
- Psychotherapy
- Thyroid
- Transcranial magnetic stimulation (TMS)

Large Effect Size

- Electroconvulsive therapy (ECT)
- Ketamines
- Light therapy
- Pramipexole
- SNT (Saint TMS)
- TMS

For High Levels of Treatment Resistance

- ECT
- High-dose monoamine oxidase inhibitors (MAOIs)

- Pramipexole
- SNT (Saint TMS)

For Prevention in Recurrent Depression

- Exercise
- Lithium
- Natural therapies for deficiency states in chapters 35–40 (possibly)
- Pramipexole
- Psychotherapy

For Anxious Depression

- Benzodiazepine augmentation
- Buspirone augmentation
- Eszopiclone augmentation
- Ketamines
- MAOI switch
- Mirtazapine augmentation
- Probiotics
- Quetiapine
- TMS
- Zuranolone (for postpartum)

For Unipolar Depression with Mixed Features

- Antidepressant taper (though bupropion and trazodone may help)
- Lamotrigine
- Lithium
- Second generation antipsychotics (especially lurasidone and lumateperone)

When the bipolar-unipolar diagnosis is unclear

- Amantadine (possibly)
- ECT
- Ketamines
- Lamotrigine (possibly)

- Light therapy
- Lithium
- Omega-3 fatty acids
- Pramipexole
- Psychotherapy
- Second generation antipsychotics
- Thyroid
- TMS

For Psychotic Depression

- Antipsychotic augmentation (high dose)
- ECT
- Lithium

For Vascular Depression

- ECT
- Nimodipine
- TMS

For Inflammatory Depression

- Bupropion
- Celecoxib
- L-methylfolate
- Minocycline
- N-acetylcysteine (2,000 mg/day)
- Omega-3 fatty acids (high dose: 4,000 mg/day)
- Pramipexole
- Probiotics
- Psychotherapy

For Depression with Trauma

- Psychotherapy
- Prazosin (Minipress) augmentation

Index

Note: This is not a comprehensive index. Page references have been limited to the most relevant discussions of each topic. **Bold page numbers** indicate the beginning of a chapter or major section where the topic is introduced. Cross-references and related terms are included where useful to help readers locate connected material efficiently.

www.ingramcontent.com/pod-product-compliance
Lightning Source LLC
Chambersburg PA
CBHW040752220326
41597CB00029BA/4734
9798999881816